Antennas with Non-Foster Matching Networks

Antennas with Non-Foster Matching Networks

James T. Aberle and Robert Loepsinger-Romak

ISBN: 978-3-031-00404-9   Paperback
ISBN: 978-3-031-00404-9   Paperback

ISBN: 978-3-031-01532-8   ebook
ISBN: 978-3-031-01532-8   ebook

DOI 10.1007/978-3-031-01532-87

Series Name: Synthesis Lectures on Antennas
Sequence in Series: Lecture #2
Series Editor and Affiliation: Constantine A. Balanis, Arizona State University
Series ISSN
Synthesis Lectures on Antennas   print 1932-6076   electronic 1932-6084

First Edition

10 9 8 7 6 5 4 3 2 1

Antennas with Non-Foster Matching Networks

James T. Aberle
Department of Electrical Engineering,
Wireless and Nanotechnology Research Center,
Arizona State University

Robert Loepsinger-Romak
MWA Intelligence, Inc.,
Scottsdale, AZ 85255, USA

SYNTHESIS LECTURES ON ANTENNAS #2

ABSTRACT

Most antenna engineers are likely to believe that antennas are one technology that is more or less impervious to the rapidly advancing semiconductor industry. However, as demonstrated in this lecture, there is a way to incorporate active components into an antenna and transform it into a new kind of radiating structure that can take advantage of the latest advances in analog circuit design. The approach for making this transformation is to make use of non-Foster circuit elements in the matching network of the antenna. By doing so, we are no longer constrained by the laws of physics that apply to passive antennas. However, we must now design and construct very touchy active circuits. This new antenna technology is now in its infancy. The contributions of this lecture are (1) to summarize the current state-of-the-art in this subject, and (2) to introduce some new theoretical and practical tools for helping us to continue the advancement of this technology.

KEYWORDS

Active antenna; electrically small antenna (ESA); non-Foster matching network

Contents

Antennas with Non-Foster Matching Networks

MOTIVATION FOR A NEW KIND OF RADIATING STRUCTURE

Anyone working in the electronics industry is aware of the trend toward increasing integration for communications and computing equipment. The holy grail of this trend is the so-called system-on-a-chip solutions. In order to fully achieve this reality, all components of the system must be capable of going on chip. Circuit design engineers have made incredible progress in developing very complex mixed-signal subsystems comprising hundreds of active devices that can fit onto a single silicon die. As a faculty member at Arizona State University, I am in awe of the amount of functionality that my analog circuit design colleagues can achieve in a tiny space on silicon. I can't help but wonder what could be achieved if somehow the same technology could be applied to antennas. However, as every decent antenna engineer knows, one critical component of radio frequency (RF) devices that does not lend itself well to integration is the antenna. Unlike digital and analog semiconductor circuits, antennas must be of a certain electrical size in order to perform their function as transducers that transform electrical signals at the input to electromagnetic waves radiating in space at the output. Certainly, I cannot be alone among antenna engineers in wondering if it is somehow possible to transform an antenna into a device that could take advantage of rapidly advancing semiconductor technology and maintain performance while dramatically shrinking in size. Indeed some preliminary steps in this direction have already been taken at Arizona State and elsewhere, and the purpose of this lecture is to summarize them and provide the necessary background for others to join the effort.

The gain-bandwidth limitation of electrically small antennas is a fundamental law of physics that limits the ability of the wireless system engineer to simultaneously reduce the antenna's footprint while increasing its bandwidth and efficiency. The limitations of electrically small antennas imply that high performance on-chip passive antennas can probably never be realized, in spite of the impact of rapidly advancing semiconductor technology on virtually all other aspects of communications systems. However, it is possible in theory to transform the antenna into an active component that is no longer limited by the gain-bandwidth-size constraints of passive antennas, and whose performance can be improved as semiconductor technology advances. This concept involves the realization of non-Foster reactive components using active circuits called negative impedance converters (NICs). These non-Foster reactances are incorporated into a matching network for the antenna that can cancel out the reactive component of the antenna's impedance and transform the radiation resistance to a reasonable value (like 50 Ω) over an octave or more of bandwidth. This revolutionary concept is only beginning to receive attention at this time. Furthermore, present technology limits the maximum frequency of non-Foster reactive components to perhaps a couple of hundred of megahertz at best. However, the potential benefits of this emerging technology are too promising to ignore. We hope in this lecture to provide the theoretical and practical framework for the future development of this exciting new technology.

The communication applications where the proposed technology would be most useful (at least initially) are likely to be low data rate, low power, short-distance, unlicensed systems. Initially, this concept is probably not going to be applicable to conventional narrowband transmit applications where active devices in the antenna would be driven into saturation by the high RF voltages present, resulting in severe distortion of the transmitted signal and concomitant severe interference at many frequencies outside of the device's assigned channel. However, for applications such as ultrawideband (UWB), RFID tags, and sensors where low transmit power is required, the construction of this type of active antenna is likely to be possible for both transmit and receive applications. This innovative approach is the key enabling technology breakthrough required for realization of completely on-chip wireless systems.

Throughout this lecture it is assumed that the reader has a sufficient background in basic antenna theory as well as analog and microwave circuit design. Excellent texts exist in both areas with the books by Balanis [1] and Pozar [2] being particularly *a propos* for this lecture. An undergraduate degree in electrical engineering is probably a minimum requirement for understanding this lecture, with a master's degree and/or several years of working experience in the area of antenna design being desirable.

ELECTRICALLY SMALL ANTENNAS

An electrically small antenna (ESA) is an antenna whose maximum physical dimension is significantly less than the free space wavelength λ_0. One widely accepted definition is that an antenna is considered an ESA at a given frequency if it fits inside the so-called *radian sphere*, or

$$k_0 a = \frac{2\pi a}{\lambda_0} < 1, \tag{1}$$

where a is the radius of the smallest sphere enclosing the antenna, $k_0 = 2\pi f/c$ is the free-space wavenumber, and $c \approx 2.998 \times 10^8$ m/s is the speed of light in vacuum. In practice, antenna

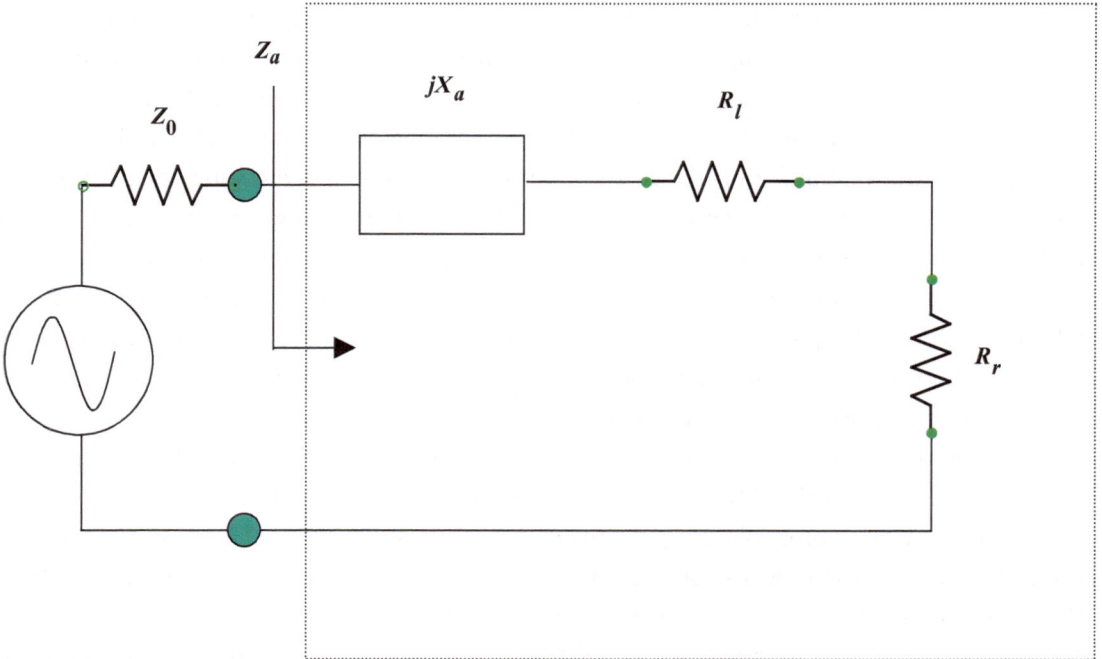

FIGURE 1: Equivalent circuit of an ESA

engineers often refer to antennas as ESAs even if they are somewhat larger than what is allowed by equation (1). In this document, we also abuse the exact definition to some extent, but assert that this does not diminish the worth of our contribution.

The input impedance of an antenna can be modeled as a lumped reactance in series with a resistance. A frequency-domain equivalent circuit for an ESA (or indeed any antenna) is shown in Fig. 1. Here R_r is the radiation resistance, which represents radiated power delivered by the antenna to its external environment, and R_l represents dissipative losses from the conductors, dielectrics, and other materials used to construct the antenna (or present in its immediate environment). For electrically small monopoles and dipoles, the reactance X_a is negative (capacitive), while for electrically small loop antenna X_a is positive (inductive). The antenna impedance is given by

$$Z_a = R_r + R_l + j X_a. \tag{2}$$

It is a common goal of antenna designers to match this (frequency dependent) impedance to some reference level (often 50 Ω) over a given bandwidth with as high efficiency as possible. The exact electrical size of the ESA determines how efficient it can be over a given bandwidth, or equivalently its gain-bandwidth product.

Theoretically, the radiation resistance of an electrically small dipole is given by

$$R_r = 20\pi^2 \left(\frac{l}{\lambda_0}\right)^2 = \frac{20\pi^2}{c^2} l^2 f^2, \tag{3}$$

where l is the physical length of the dipole (expressed in meters). For an electrically small monopole of length l, a similar equation holds:

$$R_r = 40\pi^2 \left(\frac{l}{\lambda_0}\right)^2 = \frac{40\pi^2}{c^2} l^2 f^2, \tag{4}$$

where the monopole is assumed to be mounted on an infinite perfect ground plane. (Note that for antennas with ground planes, the definition of an ESA is not so clear. One could argue that because the ground plane supports the flow of current, it is part of the radiating structure. A reasonable criterion is to declare that a monopole is an ESA if the equivalent dipole—with a length twice that of the monopole—is an ESA.)

Notice that for a fixed frequency, the radiation resistances of both dipole and monopole are proportional to the square of their length. The impedance of an electrically small loop antenna is an even stronger function of frequency with its theoretical radiation resistance given by

$$R_r = 20\pi^2 \left(\frac{C}{\lambda_0}\right)^4 = \frac{20\pi^2}{c^4} C^4 f^4, \tag{5}$$

where C is the physical circumference of the loop (expressed in meters). So the radiation resistance of the loop is proportional to its circumference raised to the fourth power. Thus, for ESAs operating at a given frequency, attempts to reduce the antenna size to fit it into a given form factor inevitably result in a dramatic reduction in radiation resistance.

One reason why this reduction in radiation resistance is undesirable can be discerned by examining the equation for the antenna's radiation efficiency. We have

$$e_{cd} = \frac{R_r}{R_r + R_l}. \tag{6}$$

From this equation, we might predict that the radiation efficiency decreases as the radiation resistance decreases. Indeed this prediction is true. But the reason this prediction is true needs further elaboration. It is not the antenna loss that is primarily responsible. (It turns out that as the antenna size is decreased, the contribution to R_l due to the antenna losses themselves also decreases albeit not as quickly as the value of R_r.) Rather, it is the losses associated with the components in the matching network that make the major contribution to the reduction in the antenna's radiation efficiency.

Another important reason to worry about the reduction of the radiation resistance is that it contributes to an increase in the antenna's radiation quality factor (Q_r). (In contrast to reactive components such as capacitors and inductors, we want an antenna to have a low quality factor.) This increase in the radiation quality factor makes the antenna more difficult (or even impossible) to match to a desired level over a given bandwidth, in accordance with a fundamental law of physics called the Bode–Fano limit.

To illustrate the concepts put forth in this lecture, we shall work with a single specific example throughout the lecture. Our example ESA comprises a cylindrical monopole mounted on an infinite ground plane. The monopole is 0.6 m in length and 0.010 m in diameter. The antenna conductor is copper but the ground plane is taken to be a perfect conductor. The frequency range of interest is around 60 MHz. (Strictly speaking this antenna is an ESA only at frequencies of 40 MHz and below. However, we allow ourselves some license here to abuse the definition as previously mentioned.) The input impedance and radiation efficiency of the monopole can be readily evaluated using a commercial software package. Here we use one called *Antenna Model*.[1] The antenna geometry as displayed in the program is shown in Fig. 2. The real part of the input impedance of the antenna obtained from the simulation is shown in Fig. 3, and the imaginary part in Fig. 4. The simulation program computes a radiation efficiency (without any matching network) of 99.8% at 60 MHz so we shall assume a radiation efficiency (before consideration of the matching network) of 100% (and hence that for the antenna by itself $R_l = 0$). It should be noted that the real part of the input impedance shown in Fig. 3 agrees quite well with the theoretical values predicted by Eq. (4), especially below about 60 MHz.

The radiation quality factor of the antenna is computed using the standard formula

$$Q_r = \frac{f}{2R_a} \left| \frac{dX_a}{df} + \left| \frac{X_a}{f} \right| \right|, \tag{7}$$

where $R_a = R_r + R_l = R_r$. A plot of the radiation quality factor for the example antenna is shown in Fig. 5. As expected for an ESA, the radiation quality factor is approximately proportional to the reciprocal of frequency to the third power. The radiation quality factor determines the bandwidth over which the antenna can be matched to a certain reflection coefficient (with an ideal lossless passive matching network), in accordance with the Bode–Fano limit to be discussed subsequently. For our example ESA, the radiation quality factor at 60 MHz is 51.9. The only way to increase the bandwidth of the ESA is to lower the total quality factor of the antenna/matching network combination by introducing loss into the matching network. The

---

[1] *Antenna Model* is available from Teri Software. It uses a method of moments algorithm based on MININEC 3, developed at Naval Ocean Systems Center by J. C. Logan and J.W. Rockway.

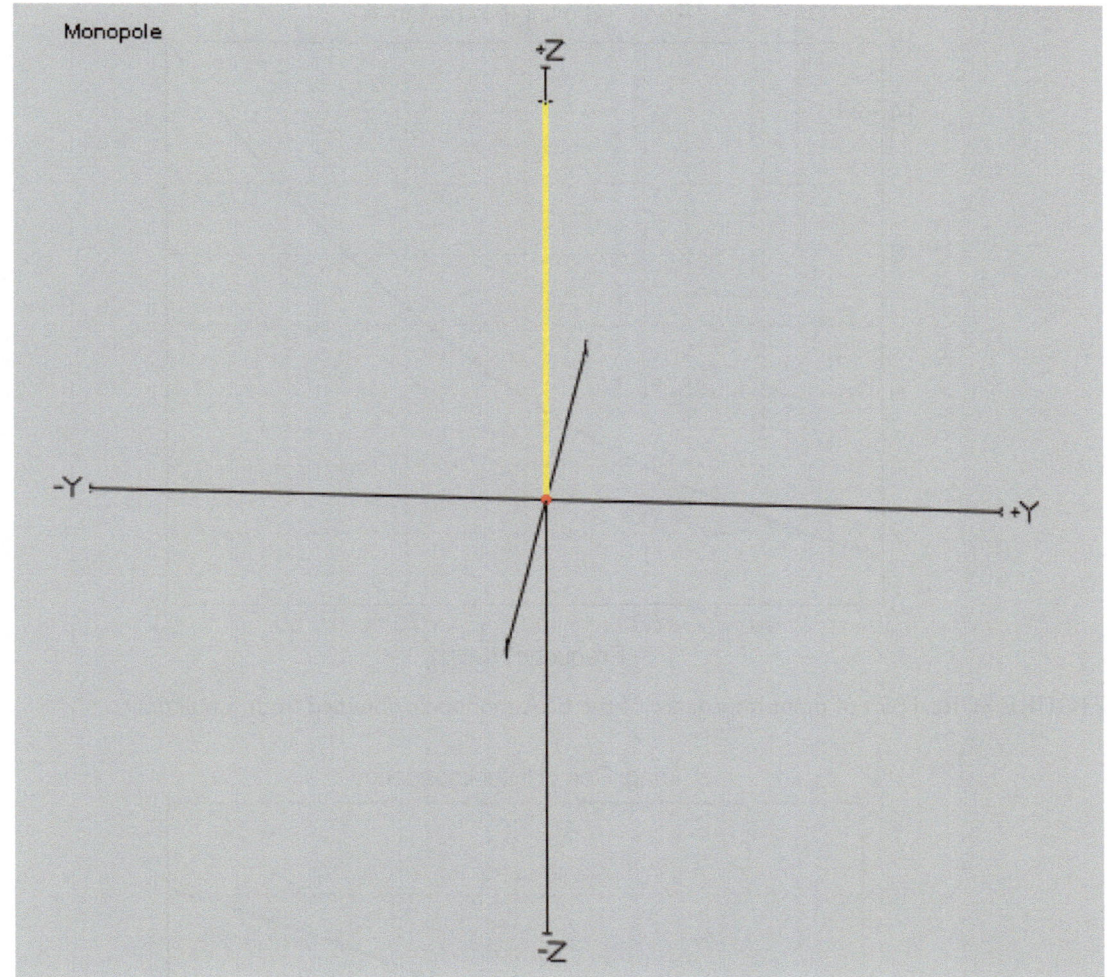

FIGURE 2: Geometry of monopole antenna as modeled in *Antenna Model* software. The monopole is a copper cylinder 0.6 m in length and 0.010 meters in diameter, mounted on an infinite perfect ground plane

total quality factor of the antenna/matching network combination is given by

$$Q_{tot} = \frac{1}{\frac{1}{Q_r} + \frac{1}{Q_m}}, \qquad (8)$$

where Q_m is the quality factor of the matching network. However, the loss in the matching network reduces the total efficiency of the system resulting in less total energy being coupled into free space.

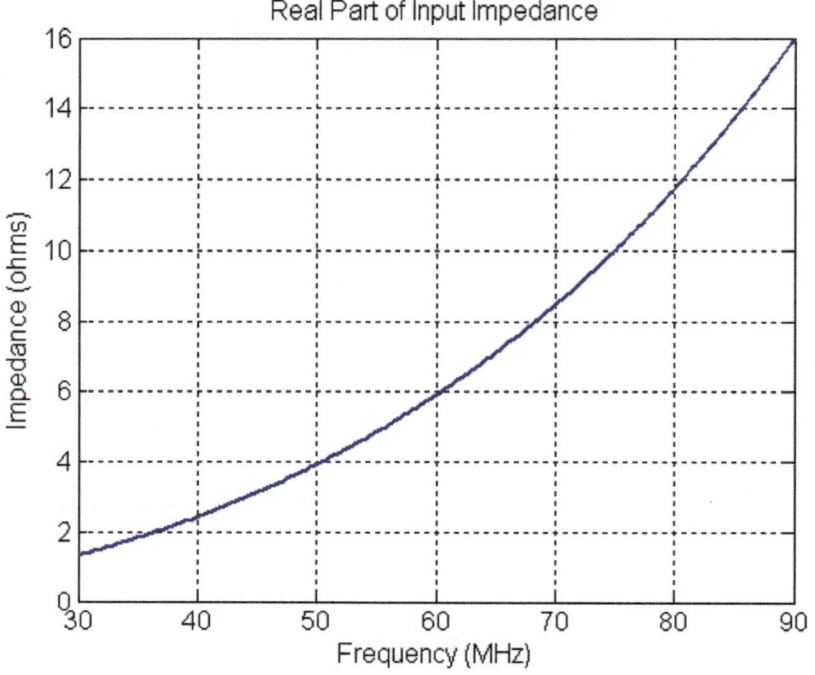

FIGURE 3: Real part of input impedance of the ESA monopole obtained from simulation

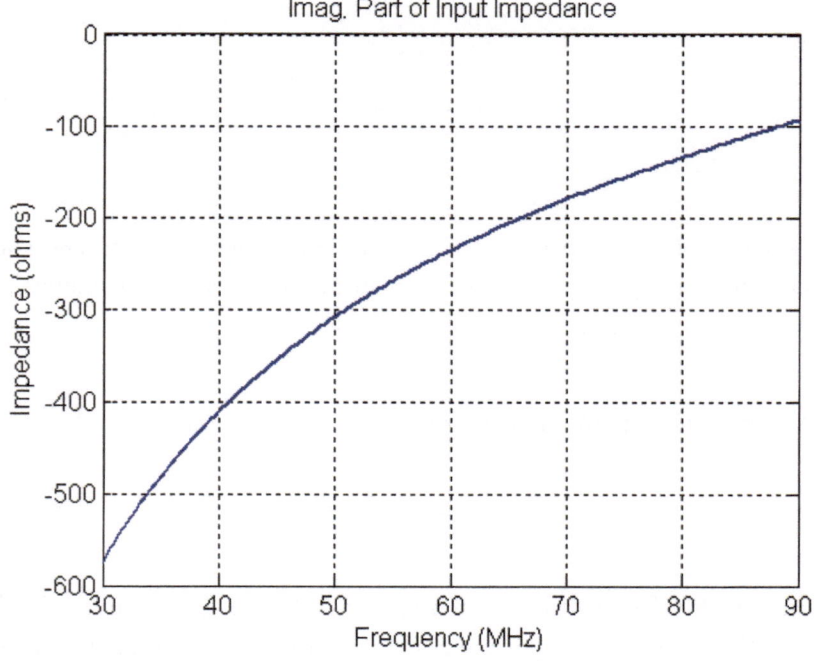

FIGURE 4: Imaginary part of input impedance of the ESA monopole obtained from simulation

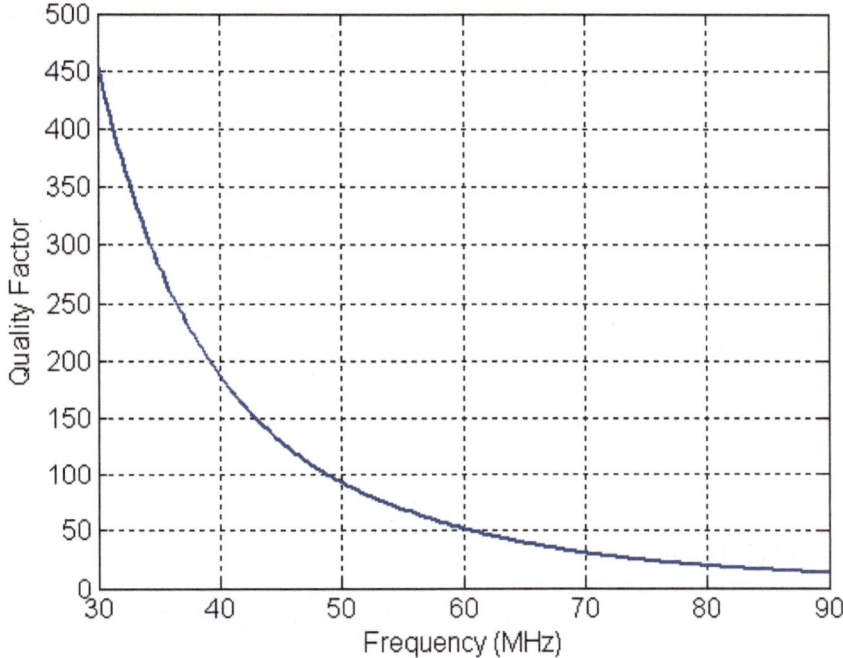

FIGURE 5: Radiation quality factor of the ESA monopole obtained from simulation

FOSTER'S REACTANCE THEOREM AND NON-FOSTER CIRCUIT ELEMENTS

Foster's reactance theorem is a consequence of conservation of energy and states that for a lossless passive two-terminal device, the slope of its reactance (and susceptance) plotted versus frequency must be strictly positive, i.e.,

$$\frac{\partial X(\omega)}{\partial \omega} > 0 \text{ and } \frac{\partial B(\omega)}{\partial \omega} > 0. \tag{9}$$

A device is called passive if it is not connected to a power supply other than the signal source. Such a two-terminal device (or one-port network) can be realized by ideal inductors, ideal capacitors, or a combination thereof.

It turns out that a corollary that follows from Foster's reactance theorem is even more important than the theorem itself. The corollary states that the poles and zeros of the reactance (and susceptance) function must alternate. By analytic continuity, we can generalize this corollary of Foster's reactance theorem to state the following about immittance (impedance and admittance) functions for a *passive* one-port network comprising lumped circuit elements:

1. The immittance function can be written as the ratio of two polynomial functions of the Laplace variable $s = \sigma + j\omega$:

$$Z(s) = \frac{N(s)}{D(s)}. \qquad (10)$$

2. The coefficients of the polynomials $N(s)$ and $D(s)$ are positive and real.
3. The difference in the orders of $N(s)$ and $D(s)$ is either zero or 1.

As two examples of the above, consider the following:

A) *Capacitor.* The impedance function is given by

$$Z(s) = \frac{1}{sC}. \qquad (11)$$

3) *Series RLC.* The impedance function is given by

$$Z(s) = R + sL + \frac{1}{sC} = \frac{s^2LC + sCR + 1}{sC}. \qquad (12)$$

If a two-terminal device has an immittance function that does not obey any of the three consequences of Foster's reactance theorem listed above, then it is called a "non-Foster" element. A non-Foster element must be an active component in the sense that it consumes energy from a power supply other than the signal source. Two canonical non-Foster elements are the negative capacitor and the negative inductor. These circuit elements violate the second consequence of Foster's reactance theorem in the list.

A) *Negative capacitor.* The impedance function of a negative capacitor of value $-C$ (with $C > 0$) is given by

$$Z(s) = \frac{-1}{sC}. \qquad (13)$$

B) *Negative inductor.* The impedance function of a negative inductor of value $-L$ (with $L > 0$) is given by

$$Z(s) = -sL. \qquad (14)$$

BASIC CONCEPTS OF MATCHING AND BODE–FANO LIMIT

It is well known from basic electrical circuit theory that maximum power transfer from a source to a load is achieved when the load is impedance matched to the source, that is when the load impedance is the complex conjugate of the source impedance. Matching between source and

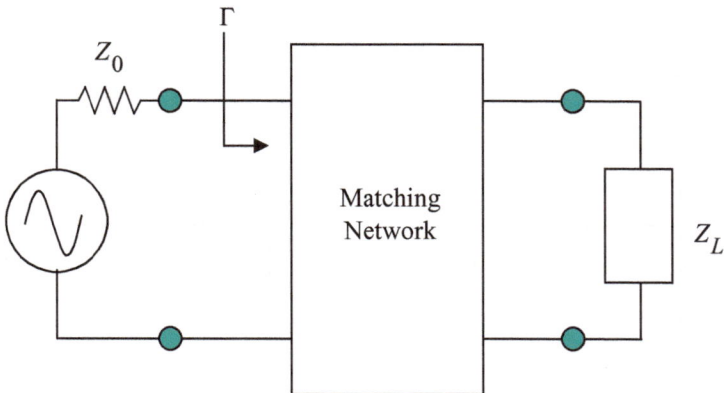

FIGURE 6: Matching network concept

load is also important so as to minimize reflections which can result in signal dispersion or even cause damage to the source. In general the load impedance is not the same as the source impedance, and a matching network is required to provide a match between the two impedances. The basic matching network concept is illustrated in Fig. 6.

Ideally, a matching network would be lossless and provide a match between the source and load over all frequencies. This is theoretically possible only if both the source and load impedances are real and the matching network is an ideal transformer. In most situations, the source impedance is real (often 50 Ω) and the load impedance is a complex quantity which varies with frequency. As a result, it is impossible to achieve an exact match (using a passive matching network) except at a single frequency (or more generally at a finite number of discrete frequencies), and the match quality degrades as frequency deviates away from this frequency. The measure of match quality is the reflection coefficient at the input of the matching network. Most commonly, the value of the reflection coefficient is represented in terms of return loss in decibels (dB). Return loss in dB is defined as

$$ \text{RL} = -20 \log_{10}(\Gamma), \tag{15} $$

where Γ is the reflection coefficient at the input of the matching network. Typically, return loss values of greater than 10 dB are considered acceptable.

The Bode–Fano criterion provides us with a theoretical limit on the maximum bandwidth that can be achieved over which a lossless passive matching network can provide a specified maximum reflection coefficient given the quality factor of the load to be matched. It should be noted that in practice a given matching network will usually provide a bandwidth that is significantly less than the maximum possible bandwidth predicted by the Bode–Fano criterion.

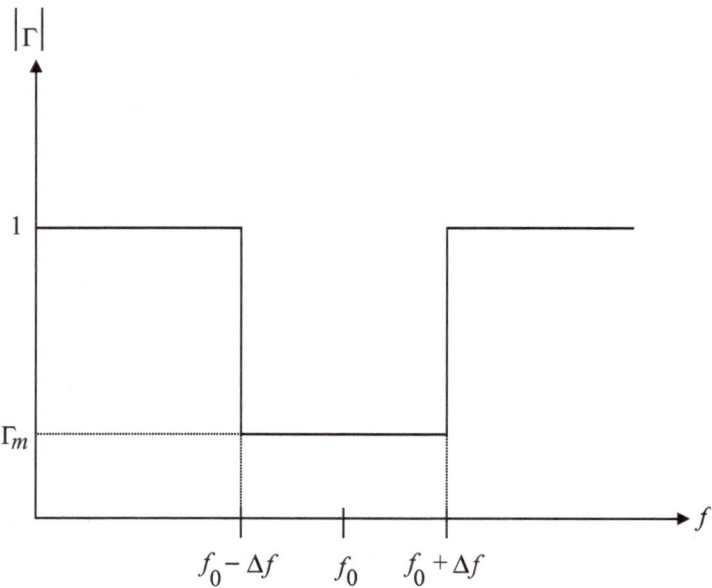

FIGURE 7: Idealized reflection coefficient response for applying Bode–Fano criterion

The most useful form of the Bode–Fano criterion may be stated as

$$\frac{\Delta f}{f_0} \leq \frac{\pi}{Q_0 \cdot \ln\left(\frac{1}{\Gamma_m}\right)}, \tag{16}$$

where f_0 is the center frequency of the match, Δf is the frequency range of the match, Q_0 is the quality factor of the load at f_0, and Γ_m is the maximum reflection coefficient within the frequency range of the match. Equation (16) is derived assuming the reflection coefficient versus frequency response shown in Fig. 7, and that the fractional bandwidth of the match is small, i.e., $\Delta f \ll f_0$. For our example ESA, Eq. (16) predicts that the fractional half-power bandwidth ($\Gamma \leq \Gamma_m = 0.7071$) achievable at 60 MHz with an ideal passive matching network is 0.042 80. This fractional bandwidth corresponds to an absolute bandwidth of about 2.6 MHz at a center frequency of 60 MHz

TWO-PORT MODEL OF AN ANTENNA

In many situations it is desirable to model an antenna as a two-port network. Such a model can be used in circuit simulations to compute the overall efficiency of the antenna with a lossy passive matching network, as well as for evaluating the stability of the network that results when an antenna is connected to a matching circuit containing non-Foster elements. The model discussed here can be applied to any antenna (with a single feed point) and does not require the assumption of a particular equivalent circuit.

Given the input impedance and radiation efficiency of an antenna at a specified frequency (from either simulation or measurements), a two-port representation of the antenna can be derived as follows. Let the complex input impedance of the antenna be denoted by Z_a, and the radiation efficiency (as a dimensionless quantity between 0 and 1) be denoted by e_{cd}. Then, we have the equivalent circuit (valid at that specific frequency) shown in Fig. 1 where

$$
\begin{aligned}
Z_a &= R_a + jX_a = R_r + R_l + jX_a \\
R_r &= e_{cd}R_a = \text{radiation resistance} \\
R_l &= (1 - e_{cd})R_a = \text{dissipative loss resistance} \\
X_a &= \text{antenna reactance.}
\end{aligned}
\tag{17}
$$

Since the radiation resistance represents power that is "delivered" by the antenna to the rest of the universe, we replace the radiation resistance with a transformer to the impedance of free space, or more conveniently, to any port impedance that we wish (such as 50 Ω). The turns-ratio of the transformer is given by

$$
N = \sqrt{\frac{R_r}{Z_0}},
\tag{18}
$$

where Z_0 is the desired port impedance. The resulting two-port representation of the antenna is shown in Fig. 8.

At each frequency, a two-port representation of the form shown in Fig. 8 can be constructed, and the two-port scattering matrix evaluated and written into an appropriate file format (such as Touchstone) for use in a circuit simulator. Note that when port 2 of the two-port shown in Fig. 8 is terminated in the proper port impedance, the antenna's input impedance is obtained as

$$
Z_a = Z_0 \frac{1 + S_{11}}{1 - S_{11}}
\tag{19}
$$

and its total efficiency is obtained as

$$
e_{\text{tot}} = |S_{21}| = \left(1 - |S_{11}|^2\right) e_{cd}.
\tag{20}
$$

In some situations, it may not be practical (or even possible) to determine the radiation efficiency of the antenna. In this case, we can usually assume a radiation efficiency of 100% (as we have done for our example ESA). Despite this assumption, the proposed model still allows us to design

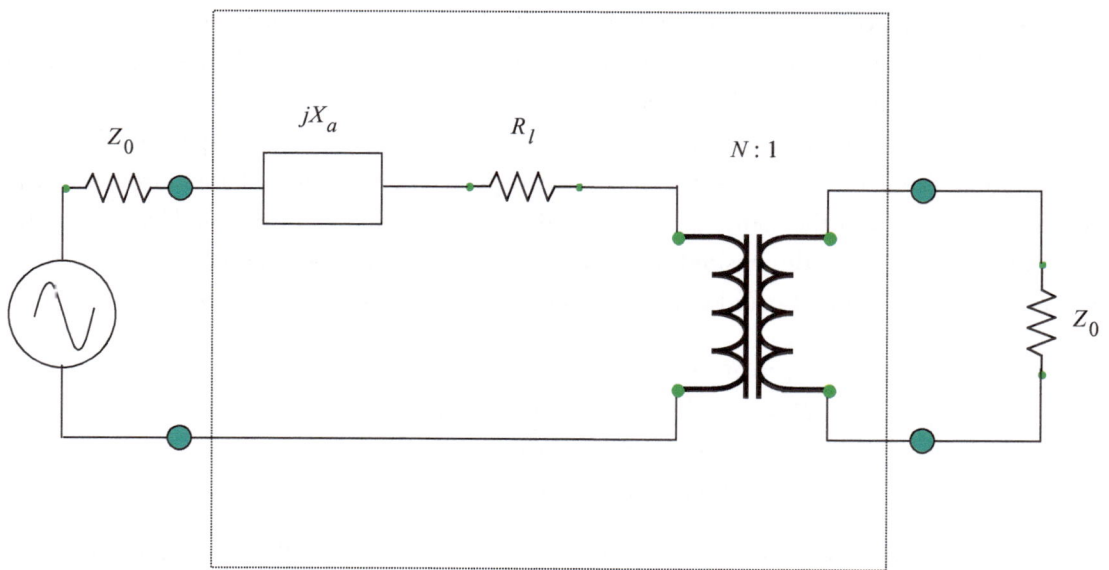

FIGURE 8: Two-port representation of an antenna (valid at a single frequency)

a matching circuit with the advantage of monitoring and optimizing both return (match), insertion loss (total efficiency), and (in the case of an active matching network) the stability of the overall circuit.

PERFORMANCE OF ESA WITH TRADITIONAL PASSIVE MATCHING NETWORK

Any number of passive matching circuits can be used to provide a (theoretically) perfect match to our example ESA at 60 MHz. One of the most common ways to match such an antenna is to use an *L*-section consisting of two inductors as shown in Fig. 9. Using readily available design formulas for the *L*-section (e.g., from Chapter 5 of [2]), one obtains the following values for the inductors when designing for a perfect match at 60 MHz:

$$L_1 = 477 \text{ nH}$$
$$L_2 = 51.9 \text{ nH}. \tag{21}$$

The major disadvantage of using a passive matching network with an electrically small antenna is that any dissipative losses in the components of the matching network reduce the overall radiation efficiency. To examine this effect, let's assume that each inductor has a Q of 100 at 60 MHz, which is reasonable for these inductance values in this frequency range. The combination of the matching network and two-port model of the antenna can be analyzed using an appropriate circuit simulator. Here we use Agilent advanced design system (ADS).

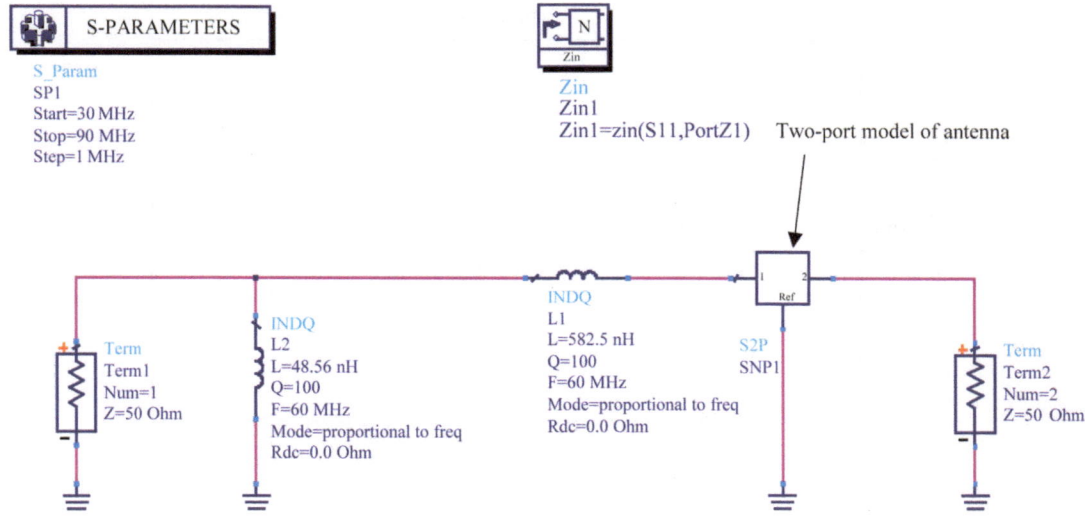

FIGURE 9: Schematic captured from Agilent ADS of ESA monopole with passive matching network

The schematic of the antenna and its matching network captured from Agilent ADS is shown in Fig. 9. The computed return loss looking into the input of the matching network is shown in Fig. 10, and the total efficiency of the antenna/matching network combination is shown in Fig. 11. Of course, the return loss result could have been obtained readily without the proposed two-port model of the antenna. However, without the use of a rigorous two-port model of the antenna, the total efficiency result would have to be calculated outside of the circuit simulator. With the use of the two-port model for the antenna, it becomes possible, for example, to use the circuit simulator's built-in optimization tools to maximize the overall radiation efficiency over commercially available inductor values, or to examine the effect of component tolerances using Monte-Carlo simulation.

As is evident from the above example, the impedance bandwidth of our example ESA with a passive matching network is quite limited. In fact, with the passive matching network shown in Fig. 9, the half power (−3 dB efficiency) bandwidth is less than 3 MHz (agreeing with our calculation using the Bode–Fano limit). As a result it is likely that any reasonable component tolerances or environmental changes would cause the antenna to be de-tuned. The antenna system's bandwidth can be increased by intentionally introducing loss into the passive matching network, but at the price of reduced maximum efficiency, the value of which can be readily evaluated inside of the circuit simulator using our approach. An interesting alternate approach that has been proposed recently is to use non-Foster reactances to provide a broadband match [3, 4].

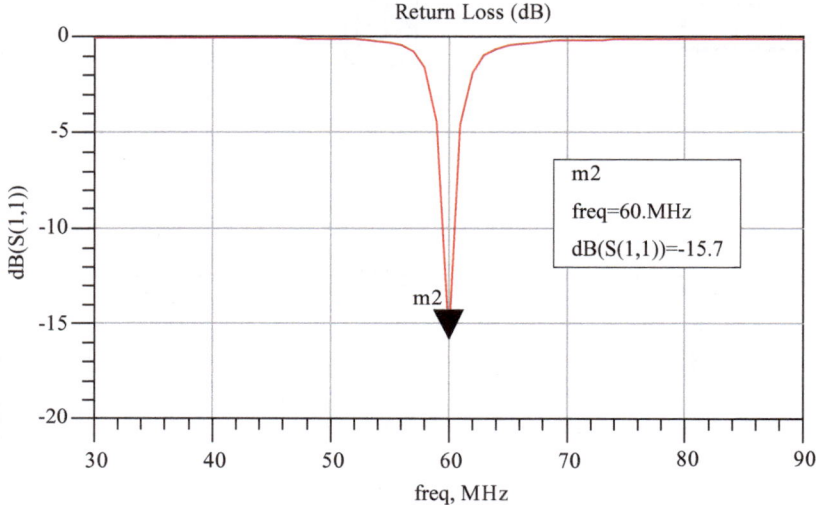

FIGURE 10: Return loss at input of passive matching network and antenna computed using Agilent ADS

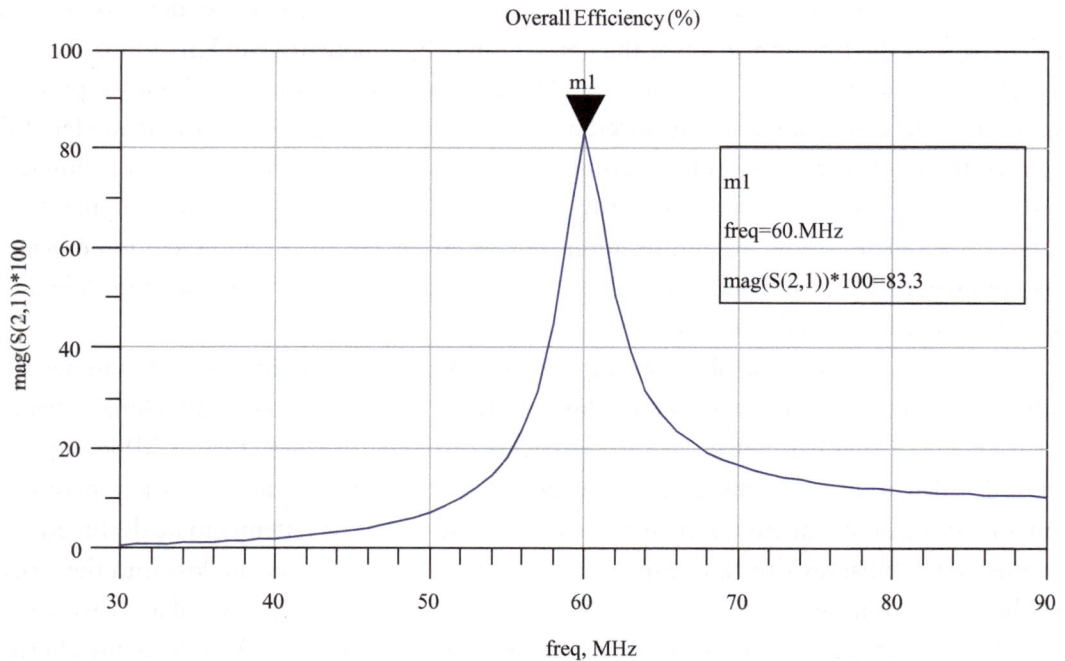

FIGURE 11: Overall efficiency (in percent) of passive matching network and antenna computed using Agilent ADS

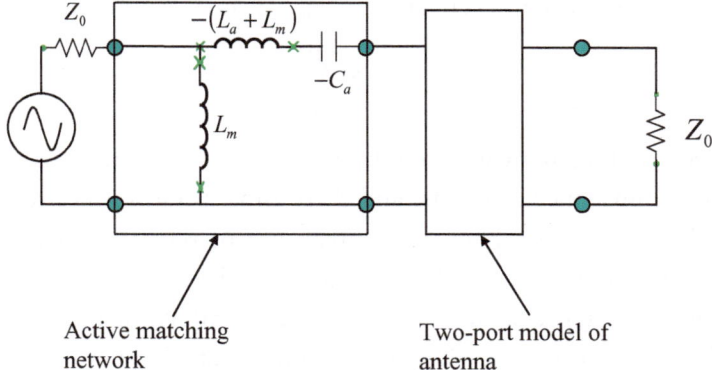

FIGURE 12: Antenna with active matching network using non-Foster reactances

PERFORMANCE OF ESA WITH IDEAL NON-FOSTER MATCHING NETWORK

A conceptual representation of the simplified ideal active matching network together with the two-port antenna model is shown conceptually in Fig. 12. The design equations for the components of the active matching network can be readily extracted from [3, 4] as follows. To design the active matching network, we first fit the antenna impedance to a simple model. Since the antenna is an electrically small monopole, the real part of the antenna impedance is assumed to vary as the square of frequency, and the imaginary part is modeled as a series LC. This simple model predicts an impedance that is denoted as $\bar{Z}_a$ and given by

$$\bar{Z}_a = R_0 \left(\frac{\omega}{\omega_0}\right)^2 + j \left(\omega L_a - \frac{1}{\omega C_a}\right). \tag{22}$$

The parameters of the model may be obtained from the "actual" antenna impedance Z_a (obtained from simulation or measurement) as

$$R_0 = Re \{Z_a (\omega_0)\}$$

$$\begin{bmatrix} \omega_1 & \dfrac{-1}{\omega_1} \\ \omega_2 & \dfrac{-1}{\omega_2} \end{bmatrix} \begin{Bmatrix} L_a \\ \dfrac{1}{C_a} \end{Bmatrix} = \begin{Bmatrix} Im \{Z_a (\omega_1)\} \\ Im \{Z_a (\omega_2)\} \end{Bmatrix}. \tag{23}$$

where ω_0 is the design frequency (in radians per second), ω_1 and ω_2 define the band of frequencies over which the model is being applied, and $Re (Z_a)$ and $Im (Z_a)$ are the real and imaginary parts of the antenna impedance respectively. The last of the necessary design equations is

$$L_m = \frac{\sqrt{R_0 Z_0}}{\omega_0}. \tag{24}$$

Basically, the active matching network works by canceling the antenna's reactance over a broadband using negative impedance elements, and then using a transformer section consisting of $-L_m$ in series and L_m in shunt to match the real part of the antenna impedance (with its frequency-squared dependence) to the desired impedance level (Z_0) over a broadband.

Using the above design equations with $\omega_1 = 2\pi \times 50$ MHz and $\omega_2 = 2\pi \times 70$ MHz, we obtain the following component values for the active matching network:

$$\begin{aligned} C_a &= 8.657 \text{ pF} \\ L_a &= 188.6 \text{ nH} \\ L_m &= 45.57 \text{ nH.} \end{aligned} \tag{25}$$

Fig. 13 shows the schematic captured from Agilent ADS of the two-port antenna model together with non-Foster matching network comprising an ideal negative inductor and capacitor. The return loss obtained from the simulation is shown in Fig. 14. Notice that the return loss is better than 10 dB from about 36 MHz to above 90 MHz, even though the antenna is electrically small. Fig. 15 shows the total efficiency of the antenna/matching network combination. Note that total efficiency better than 95% is achieved from about 36 MHz to above 90 MHz. It should also be noted that the total efficiency slightly exceeds 100% near 43 MHz. However,

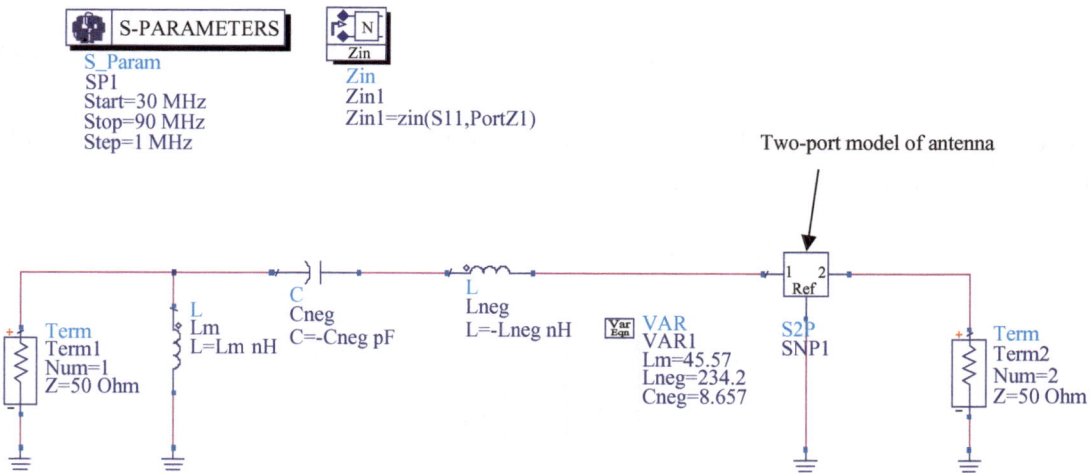

FIGURE 13: Schematic captured from Agilent ADS of ESA monopole with idealized active matching network

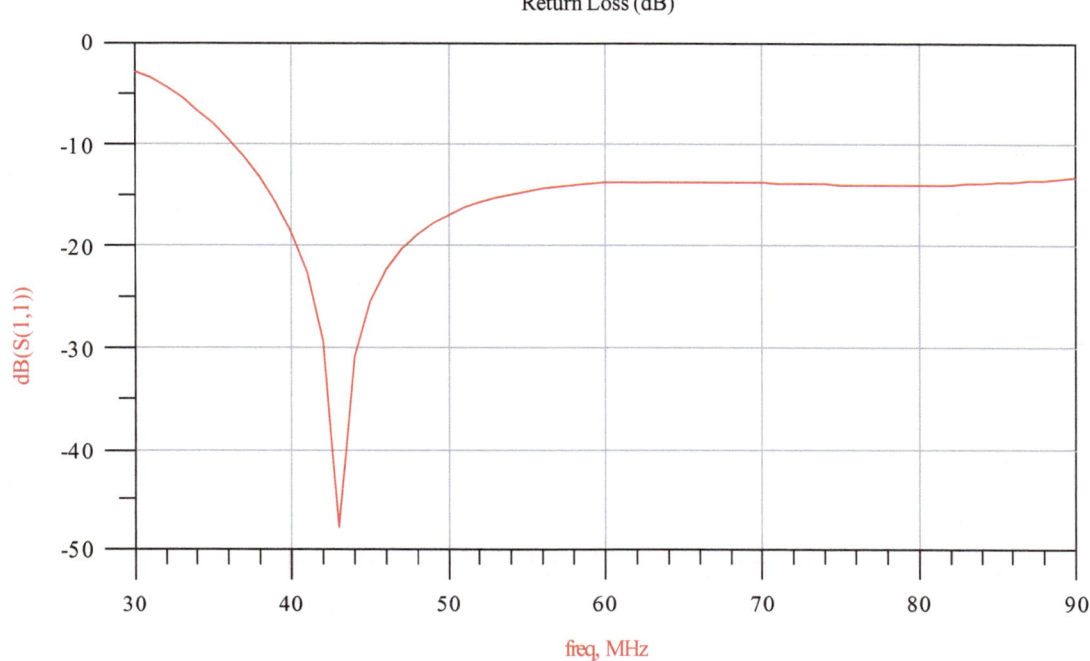

FIGURE 14: Return loss at input of idealized active matching network and antenna computed using Agilent ADS

conservation of power is not being violated because an active matching network requiring a DC power supply is implied.

Non-Foster reactances are realized using active circuits called negative impedance converters (NICs). NICs are intrinsically unstable (consider a negative resistor), and thus the stability of the combined matching network and antenna must be evaluated to ensure that the antenna does not radiate spuriously. As we shall see, the two-port antenna model allows us to readily evaluate small-signal stability measures using the circuit simulator.

BASICS OF NEGATIVE IMPEDANCE CONVERTERS (NICS)

Non-Foster behavior can be achieved by using active circuits called negative impedance converters (NICs). An ideal NIC can be defined as an active two-port device in which the impedance (or admittance) at one terminal pair is the (possibly scaled by a positive constant) negative of the impedance (or admittance) connected to the other terminal pair. An ideal NIC is shown conceptually in Fig. 16.

NICs originated in the 1920s as a means to neutralize resistive loss in circuits [5]. According-ing to Merill, negative impedance circuits were used to develop a new type of telephone repeater

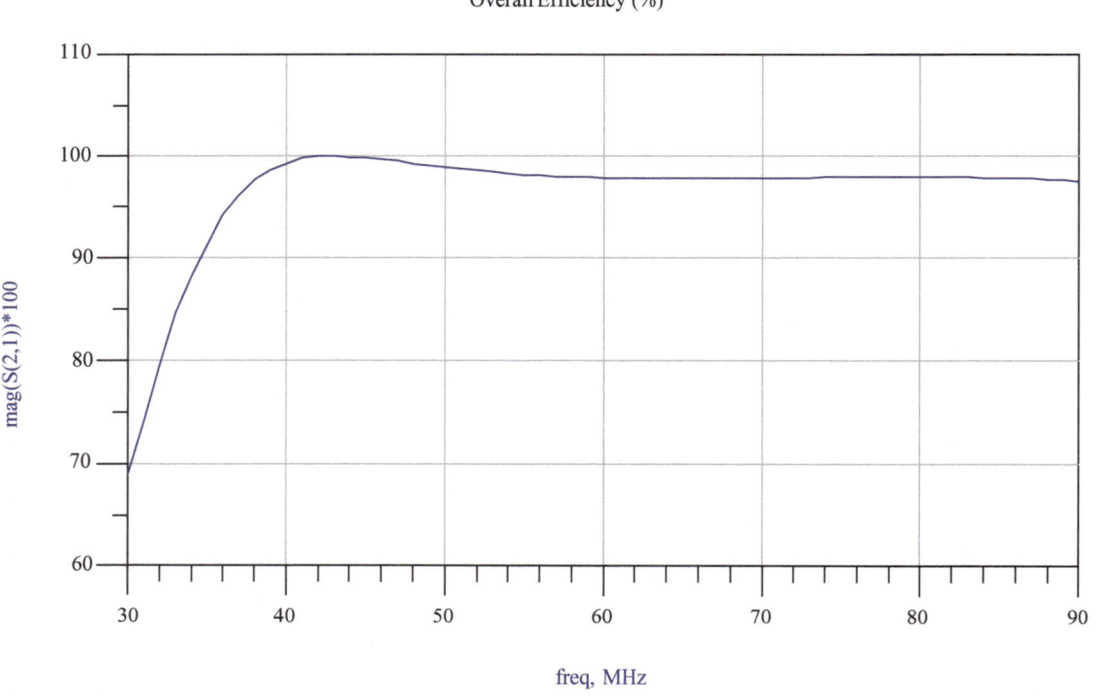

FIGURE 15: Overall efficiency (in percent) of idealized active matching network and antenna computed using Agilent ADS

called the E1. This repeater employed a feedback amplifier to provide transmission gains of 10 dB in two-wire telephone systems with extremely low loss. Due to the operation of the negative impedance circuit, the E1 repeater was able to amplify voice signals at a lower cost than conventional repeaters at the time. More recently, Yamaha incorporated negative impedance circuits in their Yamaha Servo Technology (YST) to compensate for resistive losses in the voice coil of a loudspeaker [6]. The minimization of resistive loss in the amplifier–speaker system eliminated inaccuracies in sound reproduction. Moreover, the NIC in the YST maintained

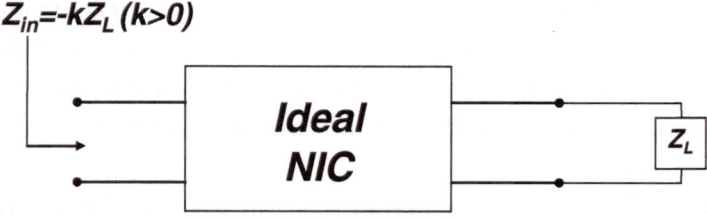

FIGURE 16: Conceptual representation of ideal NIC

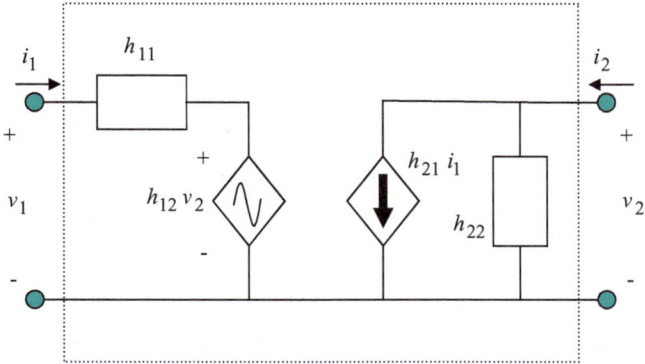

FIGURE 17: Hybrid parameter model for general two-port network

better control of the speaker cone, which allowed more air to escape through desired output ports rather than through the cone itself, resulting in maximized sound quality. Although NICs have been proven useful at audio frequencies, they have high frequency applications as well. As described in [7], a negative resistance circuit can be employed to compensate for the parasitic losses in the pass-band of a passive filter. The NIC helped to maximize the throughput (S_{21}) of a narrowband band-pass filter with a center frequency of 14 GHz.

Consider the general hybrid parameter model for a two-port network shown in Fig. 17. It is easy to show that for an ideal NIC (with $k = 1$), the following conditions must be met:

$$\begin{aligned} h_{11} &= 0 \\ h_{22} &= 0 \\ h_{12} \cdot h_{21} &= 1. \end{aligned} \quad (26)$$

Let's consider two special cases of Eq. (24): first, $h_{12} = h_{21} = -1$ and second $h_{12} = h_{21} = 1$. The first case is called a voltage inversion NIC (VINIC) since

$$\begin{aligned} v_{\text{in}} &= v_1 = -v_2 = -v_L \\ i_{\text{in}} &= i_1 = -i_2 = i_L \\ Z_{\text{in}} &= \frac{v_{\text{in}}}{i_{\text{in}}} = \frac{-v_L}{i_L} = -Z_L. \end{aligned} \quad (27)$$

The hybrid parameter model for the VINIC is shown in Fig. 18. The second case is called a current inversion NIC (CINIC) since

$$\begin{aligned} v_{\text{in}} &= v_1 = v_2 = v_L \\ i_{\text{in}} &= i_1 = i_2 = -i_L \\ Z_{\text{in}} &= \frac{v_{\text{in}}}{i_{\text{in}}} = \frac{v_L}{-i_L} = -Z_L. \end{aligned} \quad (28)$$

The hybrid parameter model for the CINIC is shown in Fig. 19.

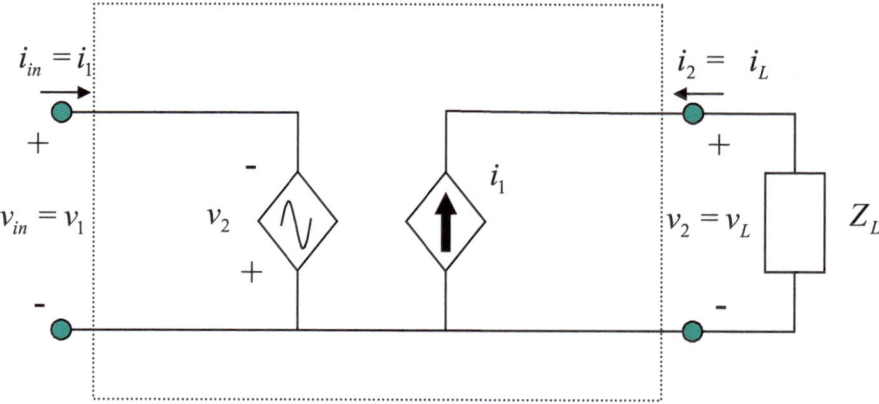

FIGURE 18: Hybrid parameter model for VINIC

The simplest practical implementation of an NIC makes use of an op-amp in the circuit shown in Fig. 20. Applying the "golden rules" of ideal op-amp analysis, we have

$$v_{\text{in}} = v_L$$
$$v_3 = v_{\text{in}} + Ri_{\text{in}} = v_L - Ri_L \Rightarrow i_{\text{in}} = -i_L. \qquad (29)$$

Thus, this simple op-amp circuit is a CINIC. Notice also that for this NIC, one side of the load is connected to ground. This type of circuit is called a grounded NIC (GNIC). The non-Foster matching circuit shown in Fig. 12 requires that the non-Foster circuit element (in that case a series negative

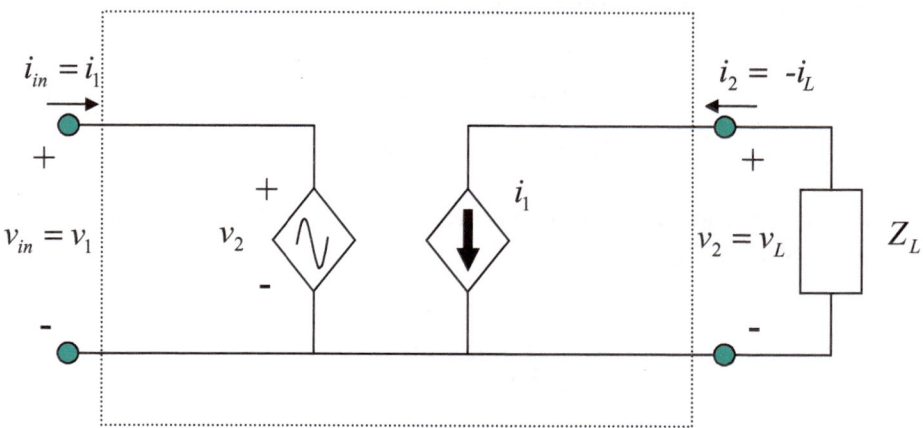

FIGURE 19: Hybrid parameter model for CINIC

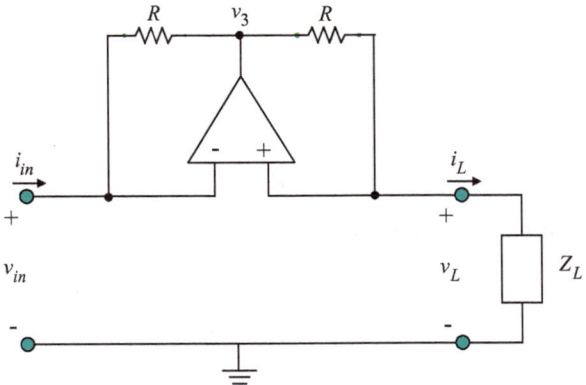

FIGURE 20: Basic op-amp NIC circuit

LC) be floating—that is, not have either side connected to ground. This type of circuit element requires what we refer to as a floating NIC (FNIC). An FNIC can be realized using two op-amps as for example in the circuit shown in Fig. 21 [8]. To demonstrate that this circuit works as an FNIC, assume that the same impedance that is to be inverted, Z_L, is also connected to port 2. If the circuit does indeed function as an FNIC, the input impedance looking into port 1 should be zero. Applying the "golden rules" of ideal op-amp analysis, we can show that

$$v_3 = v_1$$
$$v_3' = v_2$$
$$i_3 = -i_1 = i_2. \tag{30}$$

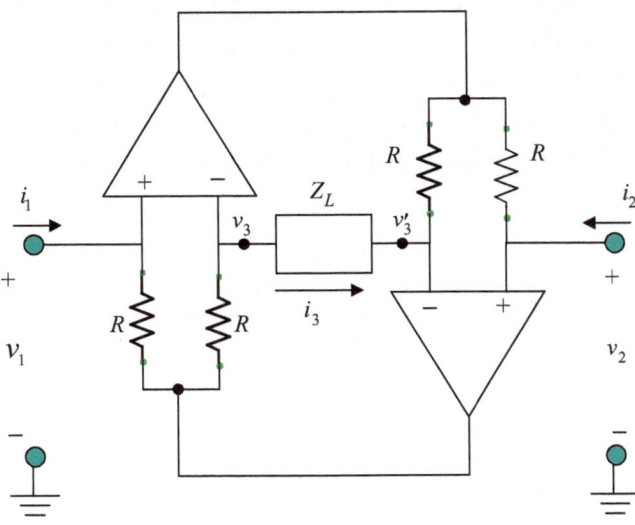

FIGURE 21: FNIC circuit using two op-amps

We also have

$$i_3 = \frac{v_3 - v_3'}{Z_L}$$

$$i_2 = -\frac{v_2}{Z_L}.$$

(31)

Combining Eqs. (30) and (31), we obtain

$$\frac{v_3 - v_3'}{Z_L} = -\frac{v_2}{Z_L}$$

or

$$\frac{v_1 - v_2}{Z_L} = -\frac{v_2}{Z_L}$$

or

$$v_1 = 0.$$

(32)

Thus,

$$Z_{\text{in}} = -\frac{v_1}{i_1} = 0$$

(33)

demonstrating that the circuit between terminals 1 and 2 acts as an FNIC. The simplified equivalent circuit of the ideal FNIC is shown in Fig. 22.

In addition to realizing NICs with op-amps, the literature contains many examples of NICs that can be realized (at least theoretically) using two transistors. In [9], a catalog of all known two-transistor NIC designs is presented. One of the earliest proposed two-transistor NICs, and the most appropriate for active matching networks since it can realize an FNIC, is shown in Fig. 23. (Note that this schematic does not show the DC biasing of the devices. The exact biasing scheme can affect circuit performance especially stability.)

To analyze the FNIC circuit shown in Fig. 23, we replace the bipolar junction transformers Q1 and Q2 with the small-signal T-model shown in Fig. 24. Doing so, we obtain the small-signal equivalent circuit for the FNIC shown in Fig. 25. To demonstrate that this circuit works as an

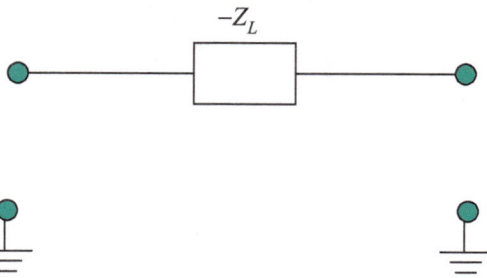

FIGURE 22: Simplified equivalent circuit of ideal FNIC

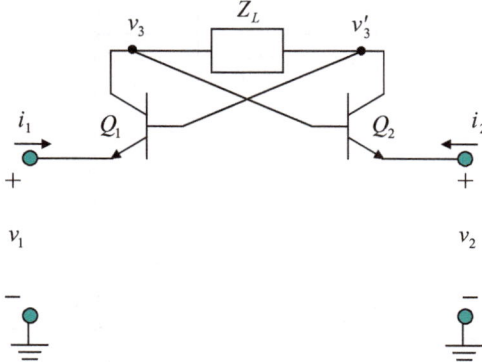

FIGURE 23: FNIC circuit using two transistors

FNIC, assume that the same impedance that is to be inverted, Z_L, is also connected to port 2. If the circuit does indeed function as an FNIC, the input impedance looking into port 1 should be zero. Utilizing nodal analysis, we can write the system of equations for the four unknown nodal voltages (v_1, v_2, v_3, and v_3') as

$$\frac{1}{r_e}v_1 - \frac{1}{r_e}v_3' = -i_1$$

$$-\left(\frac{1}{Z_L} + \frac{1}{r_e}\right)v_2 + \frac{1}{r_e}v_3 = 0$$

$$\left(\frac{1}{r_e} - g_m\right)v_1 + g_m v_2 + \left(\frac{1}{Z_L} - g_m\right)v_3 + \left(g_m - \frac{1}{r_e} - \frac{1}{Z_L}\right)v_3' = 0$$

$$g_m v_1 + \left(\frac{1}{r_e} - g_m\right)v_2 + \left(g_m - \frac{1}{r_e} - \frac{1}{Z_L}\right)v_3 + \left(\frac{1}{Z_L} - g_m\right)v_3' = 0.$$

(34)

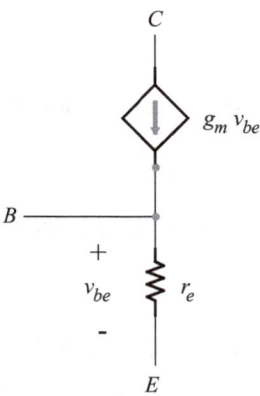

FIGURE 24: Small-signal T-model for BJT

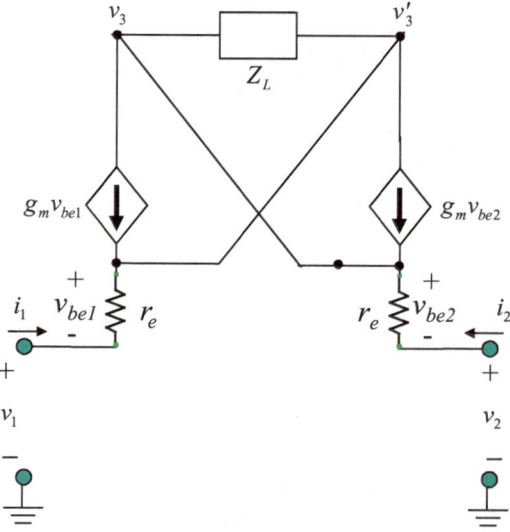

FIGURE 25: Small-signal equivalent for FNIC circuit using two transistors

Solution of the above system of equations yields

$$Z_{in} = \frac{v_1}{i_1} = 2g_m r_e Z_L - 2Z_L - 2r_e. \tag{35}$$

The general consensus in the literature seems to be that the best way (at least in theory) to realize the so-called two-transistor NICs is to replace each transistor with a kind of idealized "super transistor" called a second generation negative current conveyor (CCII-) [10]. We can think of a CCII- as a BJT with infinite transconductance (g_m). Note that for large values of transconductance, we have

$$r_e = \frac{1}{g_m}. \tag{36}$$

Hence, for an ideal transistor (with infinite transconductance), Eq. (35) yields

$$Z_{in} \xrightarrow[g_m \to \infty]{} 0. \tag{37}$$

Thus, the circuit shown in Fig. 23 behaves as an FNIC provided the transistors have large enough transconductance.

SIMULATED AND MEASURED NIC PERFORMANCE

To date we have simulated a variety of NIC circuit realizations using both small-signal S-parameter and SPICE models of the active devices. We have also constructed and measured the

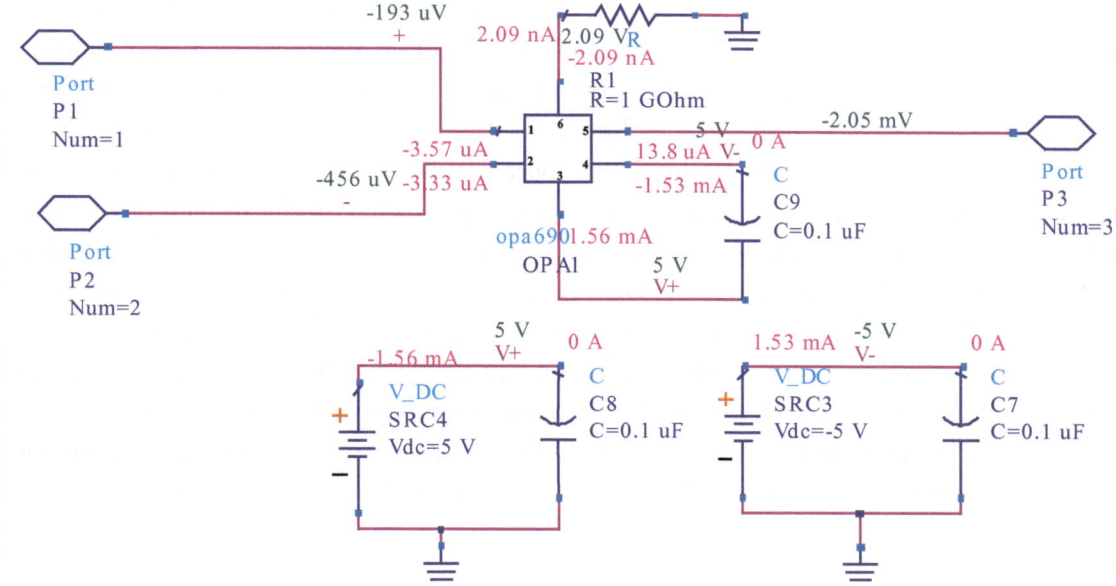

FIGURE 26: Schematic of OPA690 for simulation in Agilent ADS obtained by using the SPICE model and the data sheet for the device provided by TI

performance of several of these NIC circuits. Unfortunately, successful simulation of an NIC circuit has not always led us to a successful physical implementation. One reason for this is that all NIC circuits are only conditionally stable—that is certain auxiliary conditions must be met for the circuit to be stable. In this section we will review our progress in physically realizing NIC circuits for use in active non-Foster matching networks. The reader should be aware that this topic is one for which a great deal of work remains to be done. It is this author's opinion that the major advances in this area will be made by analog circuit designers who have been convinced by antenna engineers of the rewards to be reaped in pursuing the development of high frequency NICs.

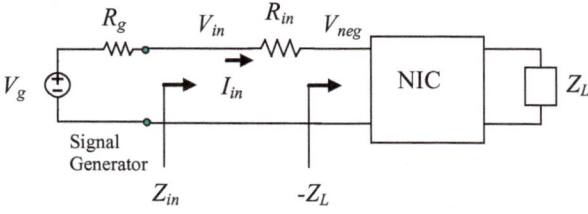

FIGURE 27: Circuit for evaluating the performance of a grounded negative impedance

The first NIC circuit that we consider is a grounded negative resistor (GNR) realized using the OPA690 op-amp from Texas Instruments (TI). The OPA690 is a wideband, voltage-feedback op-amp with a unity gain bandwidth of 500 MHz. Using the SPICE model for the device and the data sheet [11] provided by TI, an Agilent ADS model of the OPA690 can be created as shown in Fig. 26. In this circuit, port 1 is the noninverting input, port 2 is the inverting input, and port 3 is the single-ended output port. The 0.1 uF capacitors are used to RF bypass both the +5 V and −5 V power supplies, and the 1 GΩ resistor is used to simulate an open circuit for the disable pin of the OPA690 for normal operation [11]. Fig. 26 also shows the results of the DC analysis of the Agilent ADS model of the OPA690. From this analysis, we see that the overall power consumption is approximately 15.5 mW, which can be considered low power for a discrete circuit design. To characterize the behavior of the grounded negative impedance, the circuit shown in Fig. 27 is used. Fig. 28 illustrates an Agilent ADS schematic for time-domain simulation of the OPA690 GNR test circuit. The overall stability of this circuit

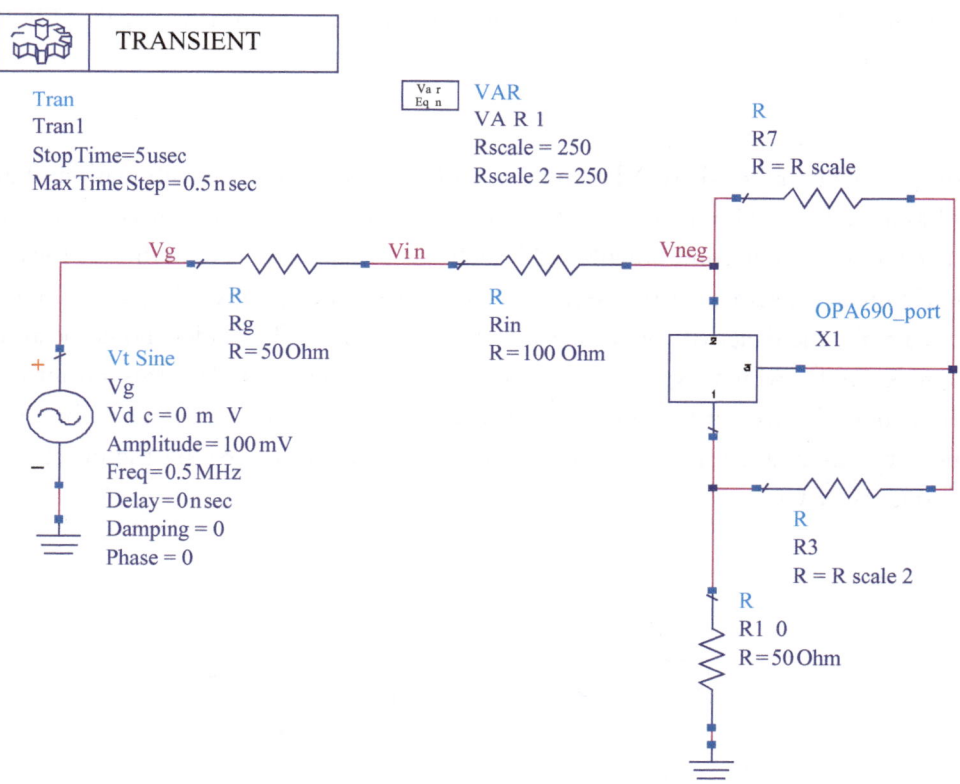

FIGURE 28: Schematic captured from Agilent ADS of the circuit for evaluating the performance of the OPA690 NIC

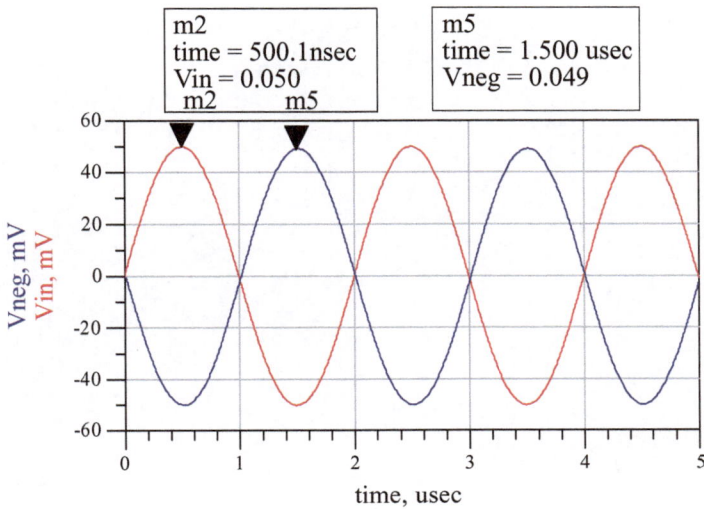

FIGURE 29: Agilent ADS simulated waveforms V_{in} and V_{neg} waveforms at 0.5 MHz for the circuit shown in Fig. 27

must be carefully considered. For high frequency, internally compensated op amps such as the OPA690, the gain as a function of frequency can be represented by [12]

$$A(s) = \frac{A_0 \omega_b}{s}, \tag{38}$$

where A_0 represents the DC gain of the op amp and ω_b represents the op amp's 3 dB frequency. Using this gain model for the op amp, the overall transfer function $T(s)$ of the OPA690 evaluation circuit (without the generator) can be computed (employing the golden rules of op-amps) as

$$T(s) = \frac{1}{\frac{Z_L - R_{in}}{Z_L + R} - \frac{s}{A_0 \omega_b}\left(1 + \frac{R_{in}}{R}\right)}. \tag{39}$$

It is well known that it is necessary for the poles of $T(s)$ to lie in the left-half of the s-plane in order for the system to be stable. Consequently, the input resistor R_{in} must be greater than the load impedance Z_L. One clever way, proposed in [9], to both ensure stability and evaluate the performance of the grounded negative impedance is to set the condition that

$$R_{in} - Z_L = 50 \ \Omega. \tag{40}$$

This choice allows us to evaluate performance in terms of return loss in a 50 Ω system using a vector network analyzer (VNA).

FIGURE 30: Photograph of fabricated OPA690 NIC evaluation board

If the GNR in the circuit of Fig. 27 is functioning properly, then ideally we should have

$$V_{neg} = -V_{in}. \tag{41}$$

Results of the time-domain simulation performed in Agilent ADS for the circuit of Fig. 28 are shown in Fig. 29. Clearly, the condition given in (41) is satisfied almost exactly and the GNR functions properly at 500 kHz.

Because of the excellent simulation results, a printed circuit board (PCB) implementation of the GNR test circuit shown in Fig. 28 was realized using readily available FR4 copper laminate and surface mount device (SMD) resistors and capacitors. Fig. 30 shows the assembled OPA690 GNR evaluation board. The simulated and measured return losses are compared in Fig. 31. In general there is excellent agreement between simulation and measurement. However, for frequencies less than 2 MHz, the measured return loss deviates somewhat from the simulation. The main cause of this discrepancy is attributed to low frequency calibration error of the VNA cables. If the 20 dB return loss bandwidth is taken to be the figure-of-merit, then the bandwidth of the OPA690 GNR is about 5 MHz. If this specification is relaxed to the 15 dB return loss bandwidth, then the bandwidth of the GNR increases to about 10 MHz. In either case, these results confirm that conventional op-amps can be used to construct NICs, but faithful negative impedance will exist only to about 10 MHz or so. The use of op-amp-based NICs at higher frequencies must await the development of op-amps with significantly higher unity gain bandwidths than are currently available. Moreover, the parasitics of the device and circuit board will have to be minimized as much as possible.

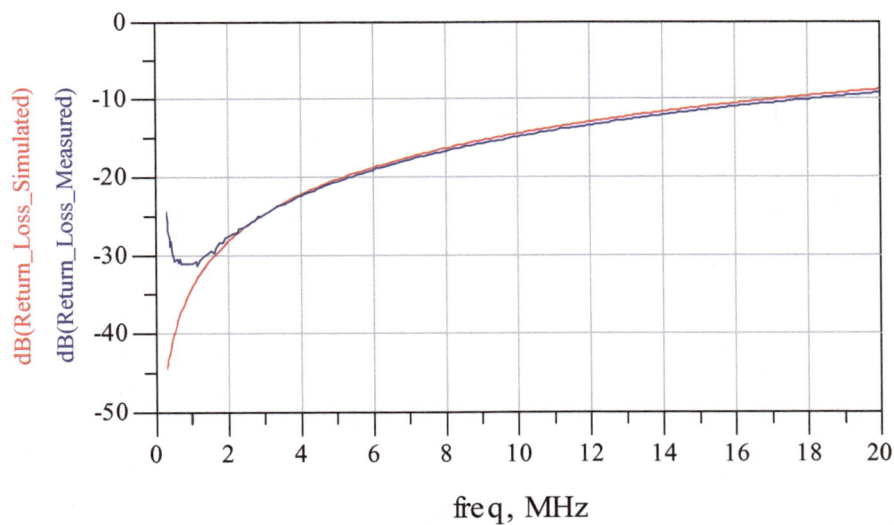

FIGURE 31: Simulated and measured return loss for the OPA690 NIC evaluation circuit

Because an op-amp's gain-bandwidth product severely limits the upper frequency at which negative impedance conversion can occur, we next focus on NIC realizations using current feedback amplifiers (CFAs) whose performance is (theoretically) not limited by their gain-bandwidth products, but mostly by their internal parasitic elements. Consequently, NICs employing these amplifiers should be more broadband in nature. To investigate this possibility, the MAX435 wideband operational transconductance amplifier (WOTA) manufactured by Maxim was selected as the NIC's active device used to realize a GNR. This device was chosen because of its simplicity, versatility, fully differential operation, and extremely wideband behavior. The current of the device is set by an external resistor R_{set} (normally 5.9 kΩ [13]), and the voltage gain of the MAX435 WOTA is set by the current gain of the device (approximately 4), the transconductance element value (Z_t), and the load resistor value (Z_L) as [13]

$$A_v = A_i \frac{Z_L}{Z_t} = 4 \frac{Z_L}{Z_t}. \tag{42}$$

This voltage gain A_v of the MAX435 was set as high as possible without its internal parasitics severely limiting the bandwidth of the amplifier. For a typical application, the load impedance Z_L must be chosen to be a finite value (usually 25 Ω or 50 Ω) [13]. A SPICE model for the MAX435 was obtained from Maxim IC's website and configured as a fully differential amplifier for simulation in Agilent ADS as shown in Fig. 32. It was found through measurement that if Z_t was less than 5 Ω, then the gain of the amplifier rolled off very quickly because a pole was introduced in the pass-band of the device. This phenomenon was modeled as an effective output

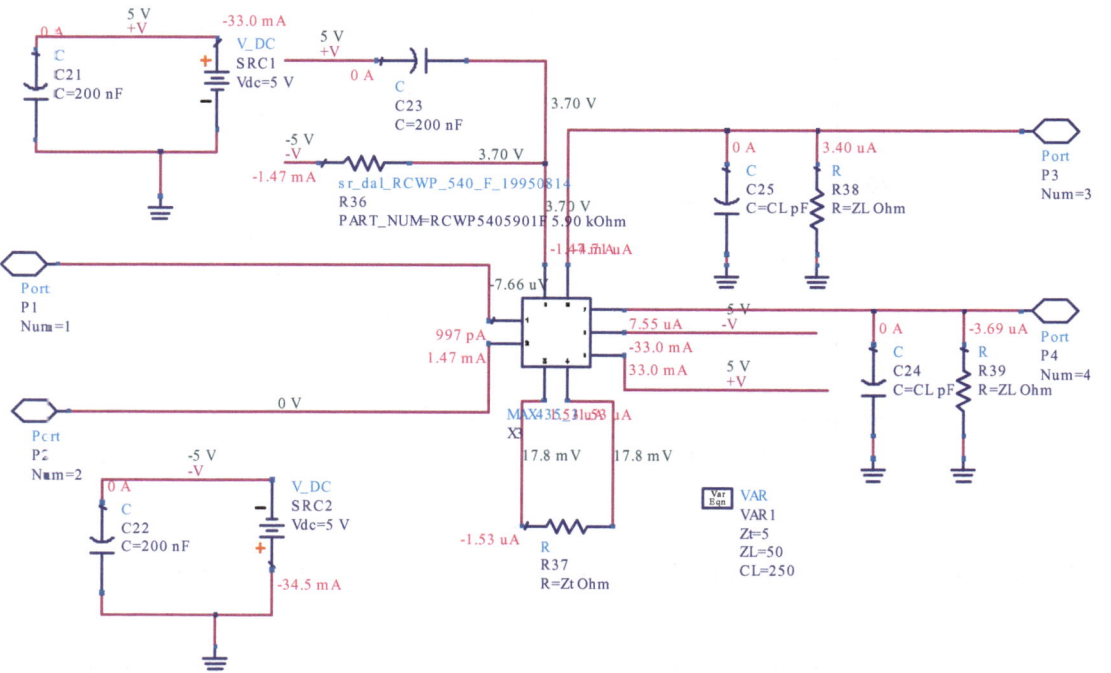

FIGURE 32: Schematic of MAX435 for simulation in Agilent ADS obtained by using the SPICE model and augmenting it to match experimental results

capacitance C_L and included in the analysis of the device. Ports 1 and 2 are the noninverting and inverting inputs, respectively, while ports 3 and 4 are the noninverting and inverting outputs, respectively. Included with the SPICE model are the external elements $Z_t, Z_L, C_L,$ and R_{set} along with power supply decoupling capacitors. The overall power consumption of the WOTA in simulation is the sum of the power of the dual supplies, which is approximately 340 mW.

Fig. 33 shows the MAX435 as a differential amplifier being used in an NIC evaluation circuit for a grounded negative resistor. The NIC topology used has been cataloged as topology *IIIa* in [6]. The MAX435 replaces both of the BJTs (or CCII-s) in the topology, thus simplifying the design and minimizing component count. Hence, a two-transistor NIC circuit can be simply constructed employing a single active device. Another distinct advantage of using the MAX435 is that no RF chokes are needed to bias the device, which allows for more compact layout schemes and reduced loss. Ideally, the input impedance of the evaluation circuit should be 50 Ω over all frequencies resulting in a reflection coefficient of zero.

As a quick proof-of-concept, the MAX435 GNR was breadboarded using a MAX435 in a 14-pin dual in-line package and surface mount discrete components. Wires with small diameters were used in some cases to create short circuits. In addition, copper tape strips were

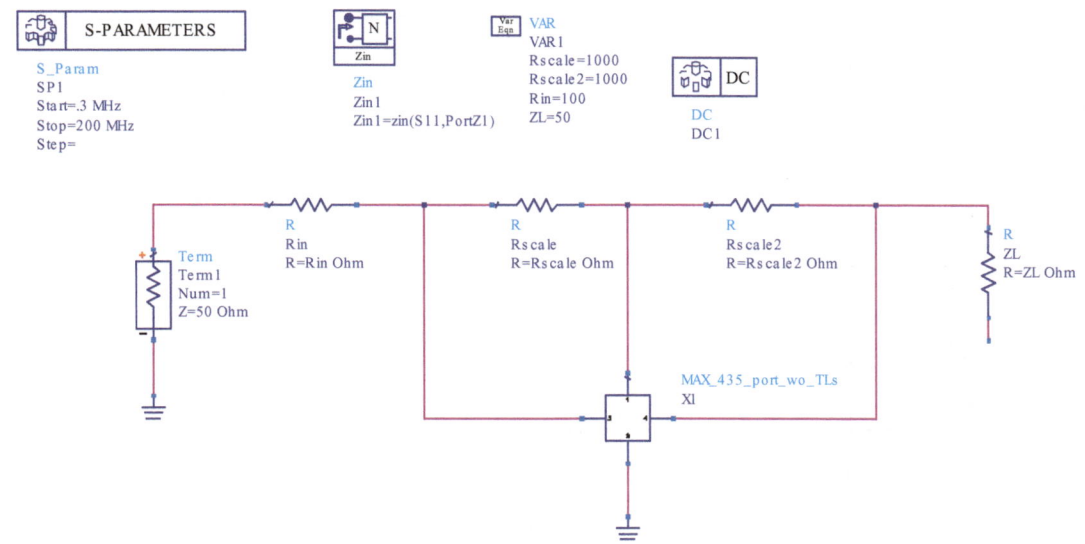

FIGURE 33: Schematic captured from Agilent ADS of the circuit for evaluation of the MAX435 NIC

used to create a good ground plane for the device as recommended in [13]. Fig. 34 shows the assembled MAX435 GNR evaluation board. The simulated and measured return losses are compared in Fig. 35. In general there is good agreement between simulation and measurement. If the 15 dB return loss bandwidth is taken to be the figure-of-merit, then the bandwidth of the MAX435 GNR is about 18 MHz.

We made a couple of unsuccessful attempts to increase the bandwidth of the MAX435 GNR circuit. In our first attempt, we replaced the MAX435 in DIP-14 package and breadboard construction with an unpackaged MAX435 and professional wirebond and PCB construction.

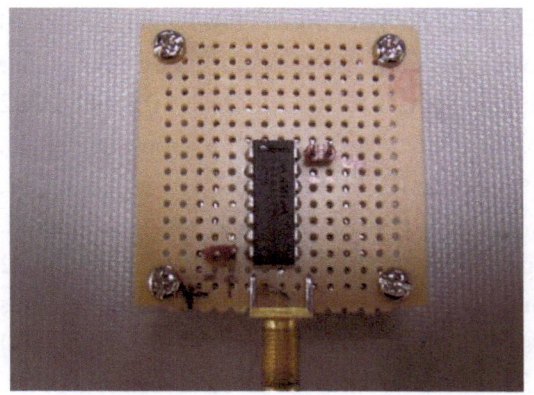

FIGURE 34: Photograph of fabricated MAX435 NIC evaluation board

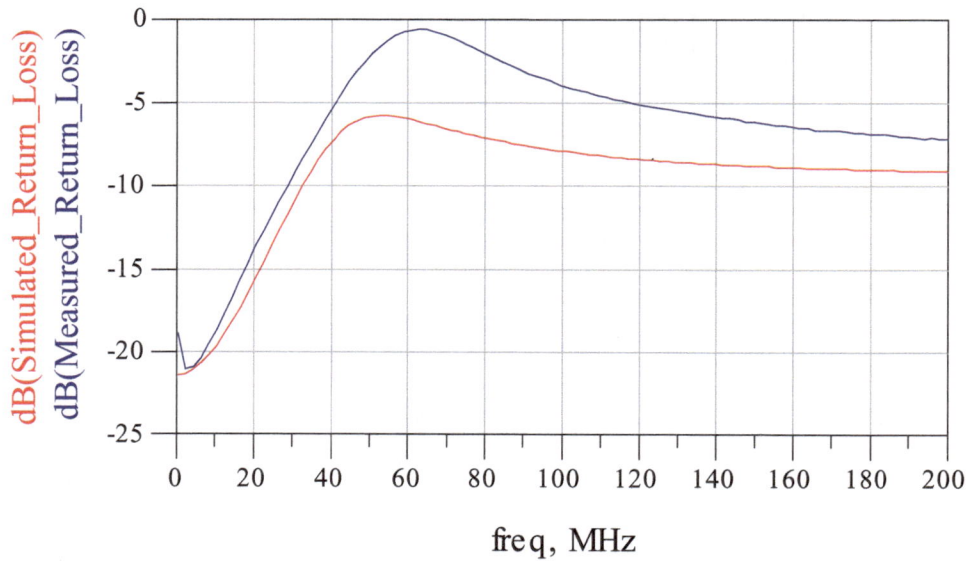

FIGURE 35: Simulated and measured return loss for the MAX435 NIC evaluation circuit

Our hope was that the new construction would greatly reduce parasitics resulting in an increase in bandwidth. Unfortunately this was not the case as the measured results for the new device were virtually identical to those of the original crude breadboard construction. In our second attempt, based on a suggestion from Maxim, we used the OPA690 as a gain-boosting stage for the WOTA. Simulations showed that this circuit should exhibit substantially improved bandwidth. Unfortunately the measured results were no better than the results we achieved with the MAX435 by itself.

The third NIC circuit considered makes use of TI's THS3202 CFA which possesses a 2 GHz unity gain bandwidth. Two amplifiers are contained within a single package. By combining the high speed of bipolar technology and all the benefits of complementary metal oxide semiconductor (CMOS) technology (low power, low noise, packing density), this amplifier is able to perform extremely well over a very large bandwidth. A SPICE model for the THS3202 can be downloaded from TI's website and was implemented in Agilent ADS as shown in Fig. 36. The inductor and capacitor form a low-pass filter to prevent AC ripple on the power supply line. The THS3202 can be configured as a GNR much like the OPA690 GNR previously considered. Following the design guidelines in [14], the scaling resistors R_{s1} and R_{s2} were chosen to be 200 Ω to maximize the gain and minimize the overall noise figure of the amplifier. Physical realizations of THS3202 GNR circuits were implemented using an evaluation module (THS3202 EVM) that was purchased through TI and shown in Fig. 37. This board was modified to realize a GNR. The simulated and measured return losses are compared in Fig. 38. If the 20 dB return loss

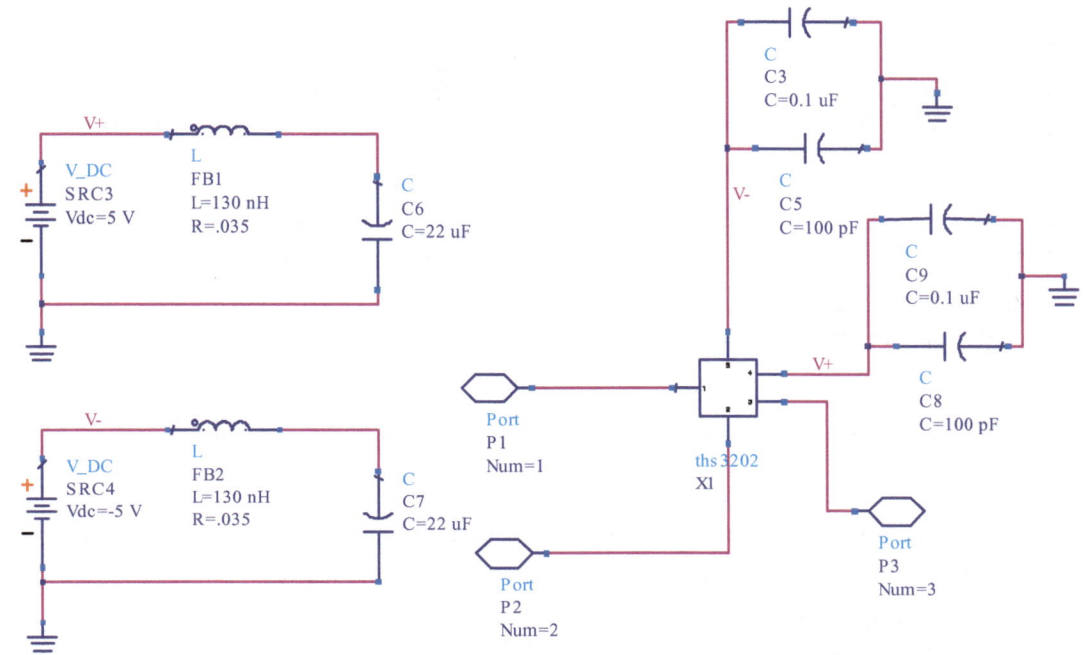

FIGURE 36: Agilent ADS model of the THS3202 with supply bypassing

bandwidth is taken to be the figure-of-merit, then the simulation bandwidth of the THS3202 negative resistor evaluation circuit is about 120 MHz. Unfortunately, the measured bandwidth is only about 50 MHz. Nevertheless, the measured results for the THS3202 GNR are still significantly greater than the results obtained using either the OPA690 or the MAX435 as the NIC's active devices. In the simulation, the measured input resistance of the THS3202 GNR

FIGURE 37: Photograph of THS3202 evaluation board (THS3202 EVM) purchased from TI and modified to form an NIC evaluation circuit

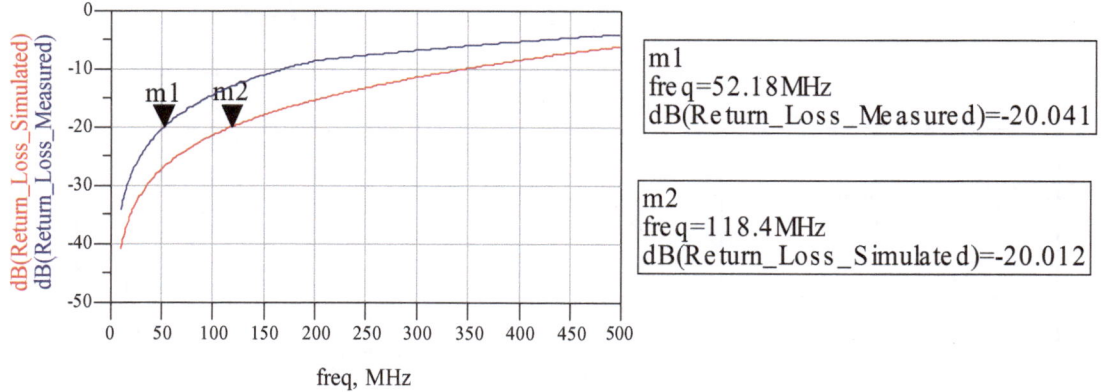

FIGURE 38: Simulated and measured return loss for the THS3202 NIC evaluation circuit

is very nearly equal to −50 Ω to frequencies greater than 500 MHz. However, the reactance of the THS3202 GNR is nonzero and behaves like a parasitic inductance. Thus, potentially we may be able to compensate for it and extend the bandwidth of the circuit.

Having had some success in fabricating GNRs, we turned our attention to floating negative resistors (FNRs). This work is still in its early stages, and only simulation results are presented here.

To implement an FNIC, two THS3202 amplifiers (in the same package) can be used to realize the circuit shown in Fig. 21. The schematic of the FNIC captured from Agilent ADS is shown in Fig. 39. As with all the NIC circuits, particular attention needs to be paid to stability. Each of the GNR circuits previously considered is a one-port device that can be stabilized by employing a series resistor R_{in} that also allowed evaluation of the overall reflection coefficient S_{11} in a 50 Ω system. The return loss of the resulting one-port was used as a figure-of-merit for the bandwidth of the GNR. To assess the performance of a floating negative impedance circuit, we can construct a so-called all-pass two-port network using the circuit shown in Fig. 40. Not only does this approach allow evaluation of the input return loss and the insertion loss as figures-of-merit, it also allows one to evaluate the small-signal stability of the network using conventional two-port measures. For the circuits that we consider here, the FNR has (ideally) an equivalent series resistance of −50 Ω that negates a series 50 Ω resistor. As a result, both the input and output impedances of the circuit should be 50 Ω. In Fig. 41, a schematic captured from Agilent ADS shows the THS3202 FNIC configured as a −50 Ω FNR and placed into an all-pass system configuration with a load impedance $R_L = 50\Omega$ across ports 3 and 4. Notice in the schematic the presence of the μ' token which allows the assessment of the small-signal stability of the network. Simulated results for return loss and small-signal stability of the THS3202 FNR in the all-pass network are shown in Fig. 42. Although the −20 dB return loss bandwidth is

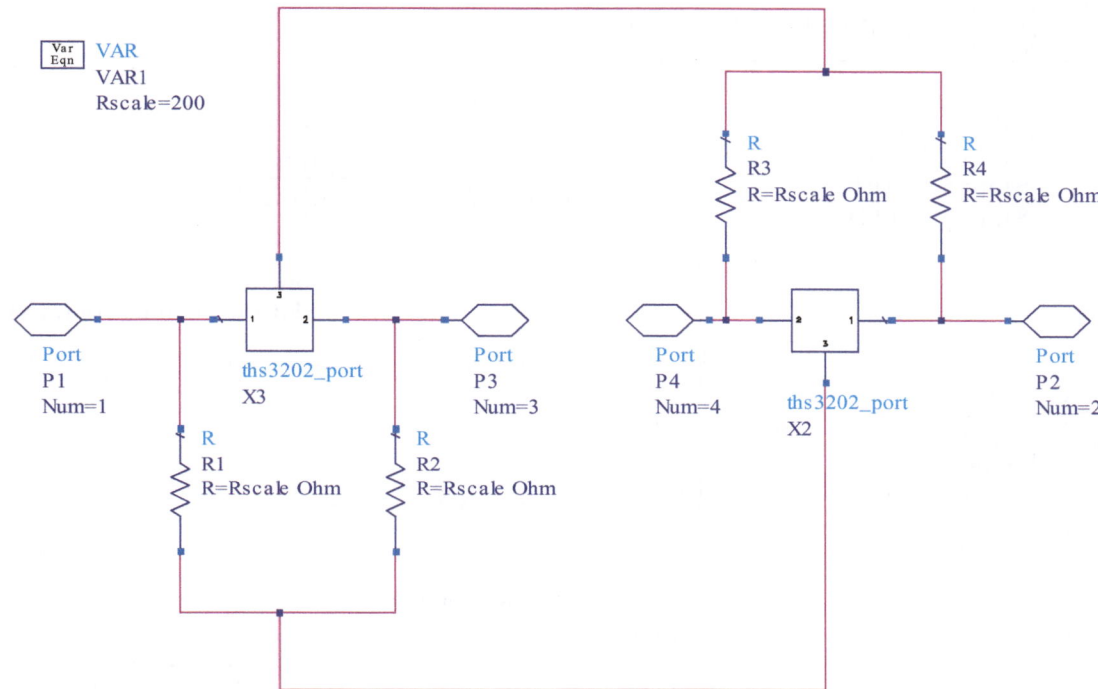

FIGURE 39: Schematic captured from Agilent ADS of the THS3202 FNIC circuit

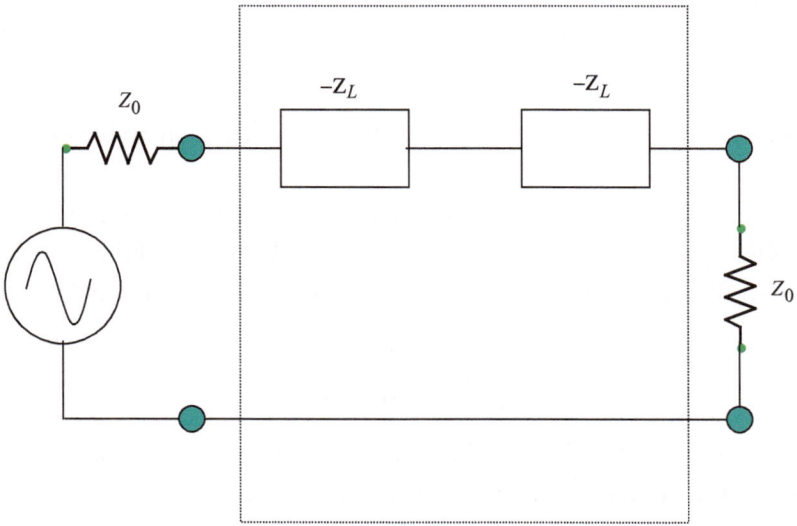

FIGURE 40: All-pass circuit for evaluating the performance of a floating negative impedance

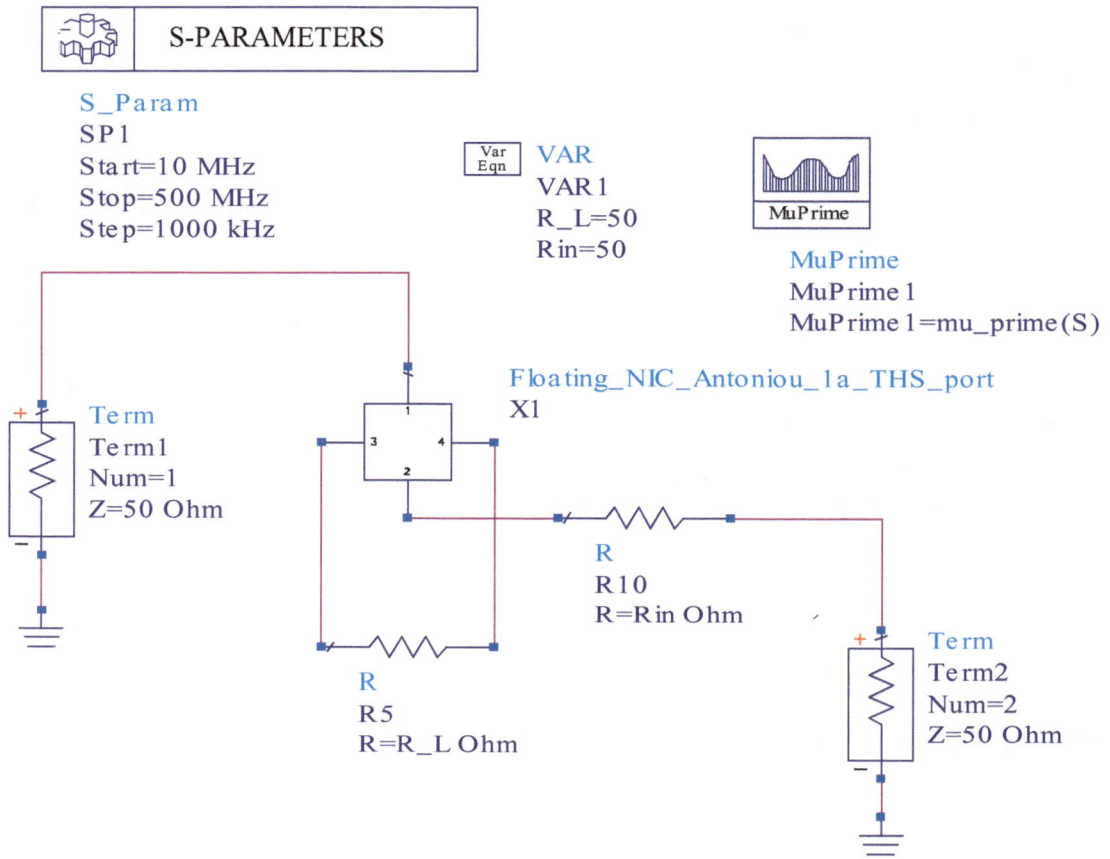

FIGURE 41: Schematic captured from Agilent ADS of the THS3202 FNIC of Fig. 38 configured as a FNR and installed in the all-pass evaluation circuit

broadband (approximately 100 MHz), the circuit is unconditionally stable only for frequencies less than 50 MHz.

In an attempt to create an FNR with greater small-signal stability, we arranged two THS3202 GNRs back-to-back as shown in Fig. 43. Analyzing the circuit assuming ideal op-amps, we find that the equivalent resistance seen between ports 1 and 2 is given by

$$R_{in} = \frac{R_3 R_1}{R_1 - R_2}. \tag{43}$$

Consequently, for the input resistance R_{in} to be the negative of the load impedance R_3, the following relationship between R_1 and R_2 must be chosen as

$$R_2 = -2R_1. \tag{44}$$

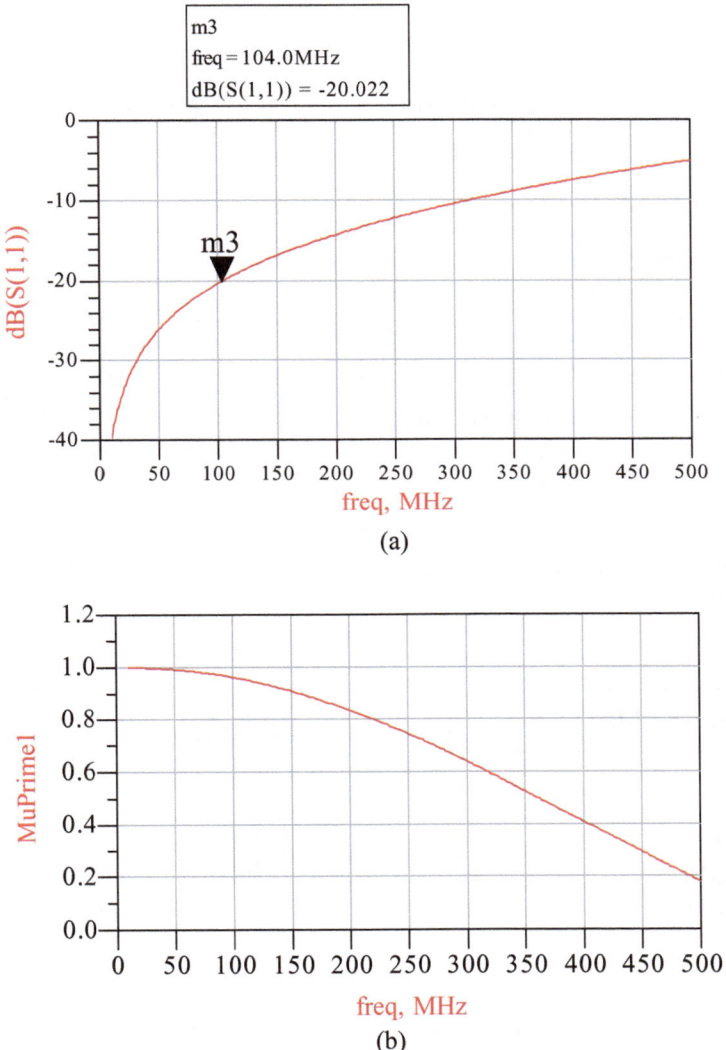

FIGURE 42: Simulated (a) return loss and (b) stability of the all-pass test circuit for the THS3202 FNR

An all-pass implementation simulation in Agilent ADS with $R_3 = 50\Omega$ is depicted in Fig. 44 where the FNR is placed inside a two-port data item box. To minimize noise and maximize gain, R_1 and R_2 are chosen to be as small as possible (200 Ω and 400 Ω, respectively) without affecting the performance of the FNR. The two 25 Ω resistors on each side of the FNR complete the all-pass test circuit. The simulation results of Fig. 45 show that the -20 dB return loss bandwidth is only about 30 MHz, but the network is close to being unconditionally stable over almost the entire frequency range. We found that the input reactance X_{in} is negative and so might be compensated over a limited frequency range using a series inductor. By trial and

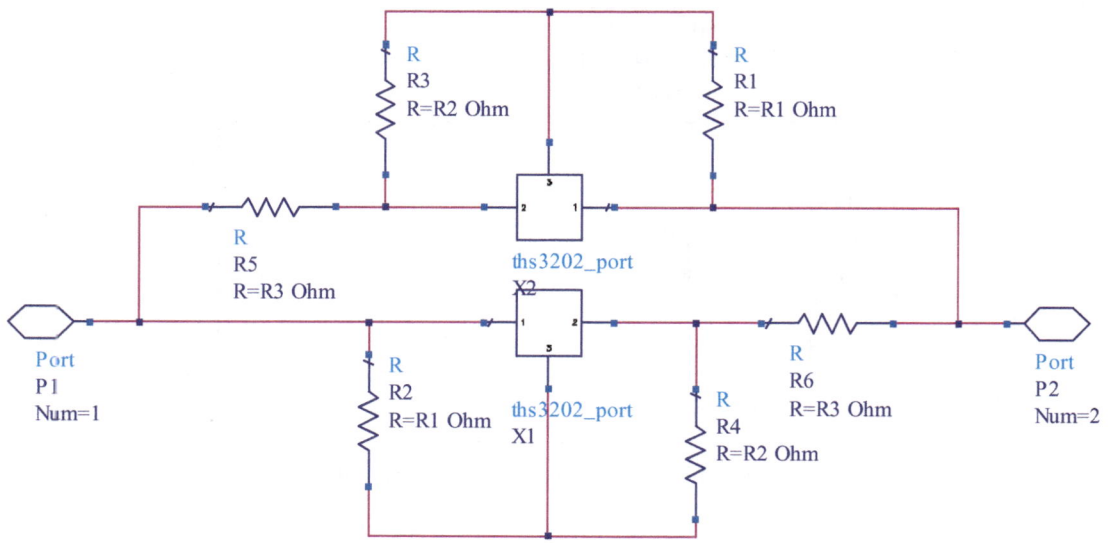

FIGURE 43: Schematic captured from Agilent ADS of the THS3202 FNR circuit formed by two back-to-back GNRs

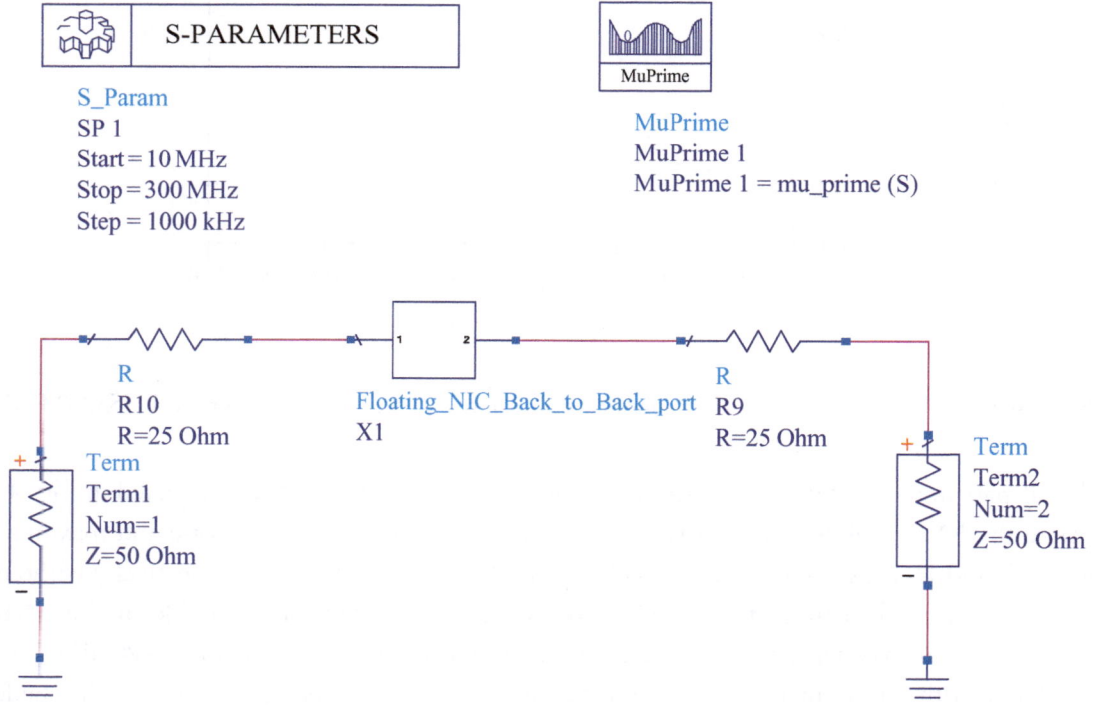

FIGURE 44: Schematic captured from Agilent ADS of the THS3202 FNR of Fig. 42 installed in the all-pass evaluation circuit

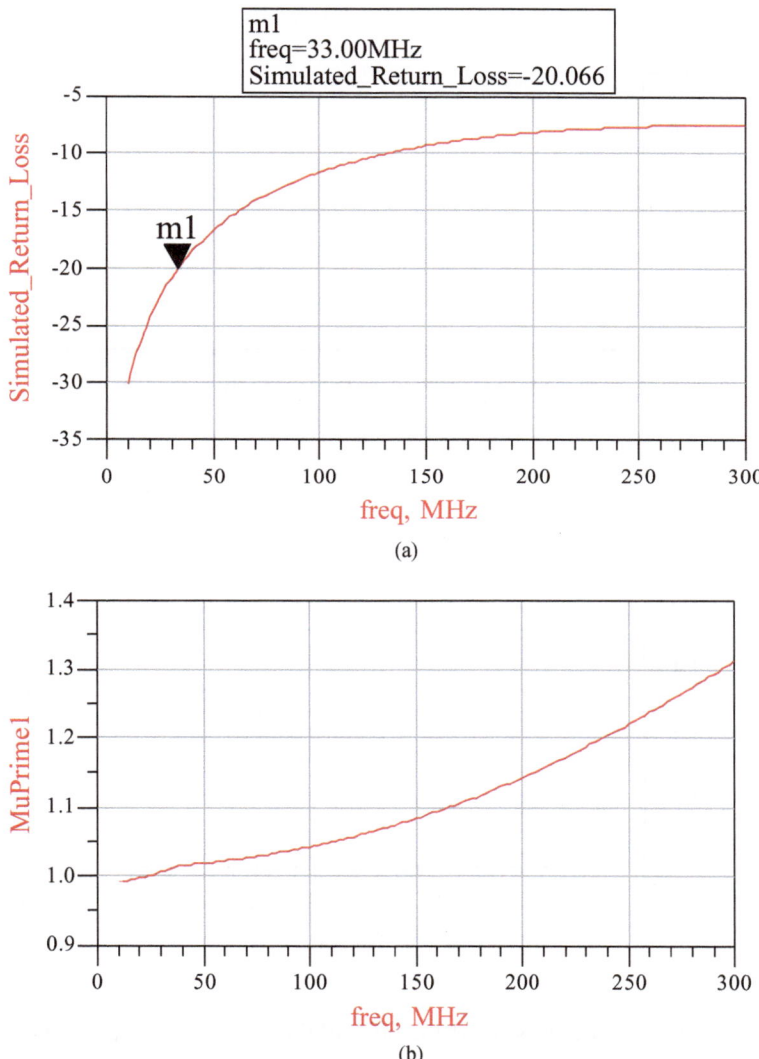

FIGURE 45: Simulated (a) return loss and (b) stability of the all-pass test circuit for the THS3202 FNR formed by two back-to-back GNRs

error, we found that placing an inductance of 45 nH in series with the FNR maximized the return loss bandwidth and stability of the all-pass test circuit as shown in Fig. 46. The simulated −15 dB return loss bandwidth is expanded to greater than 250 MHz. Unfortunately, the circuit is not unconditionally stable for frequencies less than 125 MHz, but may be relatively easy to stabilize since μ' is so close to unity.

Another way to implement an FNIC is to use two BJTs to realize the circuit shown in Fig. 23. Following the work reported in [15] and [16], we use the NE85630 NPN silicon

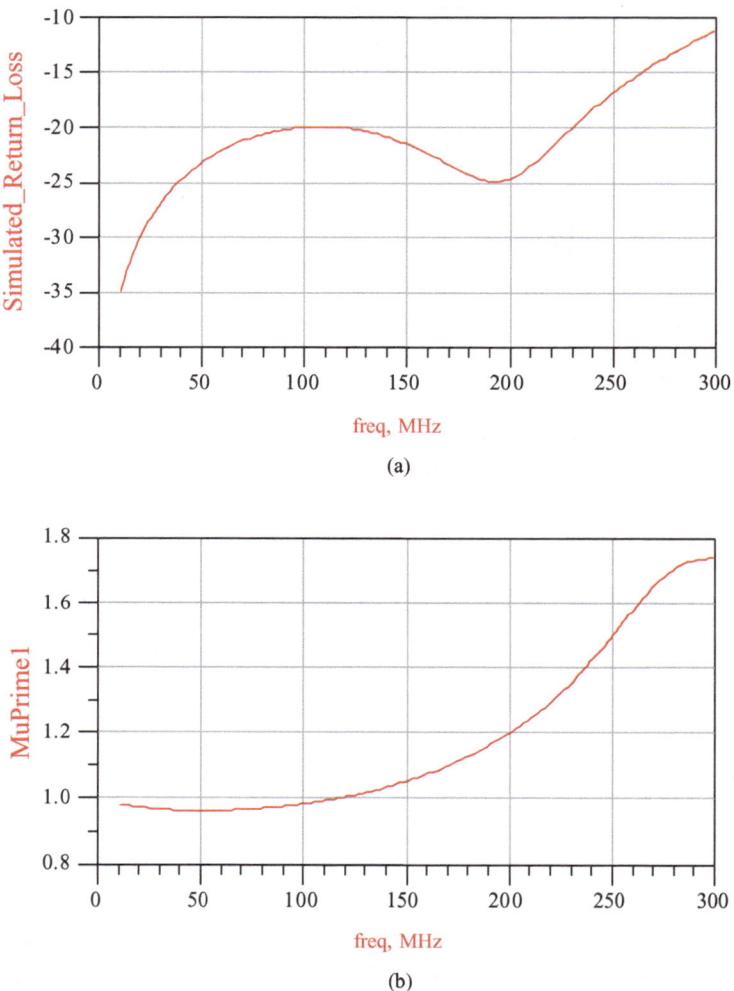

(a)

(b)

FIGURE 46: Simulated (a) return loss and (b) stability of the all-pass test circuit for the THS3202 FNR formed by two back-to-back GNRs with a 45 nH series inductor

RF transistor from NEC. The schematic of the FNR all-pass test circuit using these devices captured from Agilent ADS is shown in Fig. 47. The simulated performance of this FNR test circuit is shown in Fig. 48. As can be seen, the −20 dB return loss bandwidth approaches 200 MHz, and the circuit is unconditionally stable at all simulation frequencies. It should be noted that the simulation is performed using only the S-parameters of the NE85630 (rather than a SPICE model) valid under a specified bias condition.[2] The exact details of the biasing circuit are

---

[2]S-parameters for the NE85630 device are provided from 50 MHz to 3.6 GHz. Since we are simulating our circuits below 50 MHz, we are also relying on an accurate extrapolation of the S-parameters.

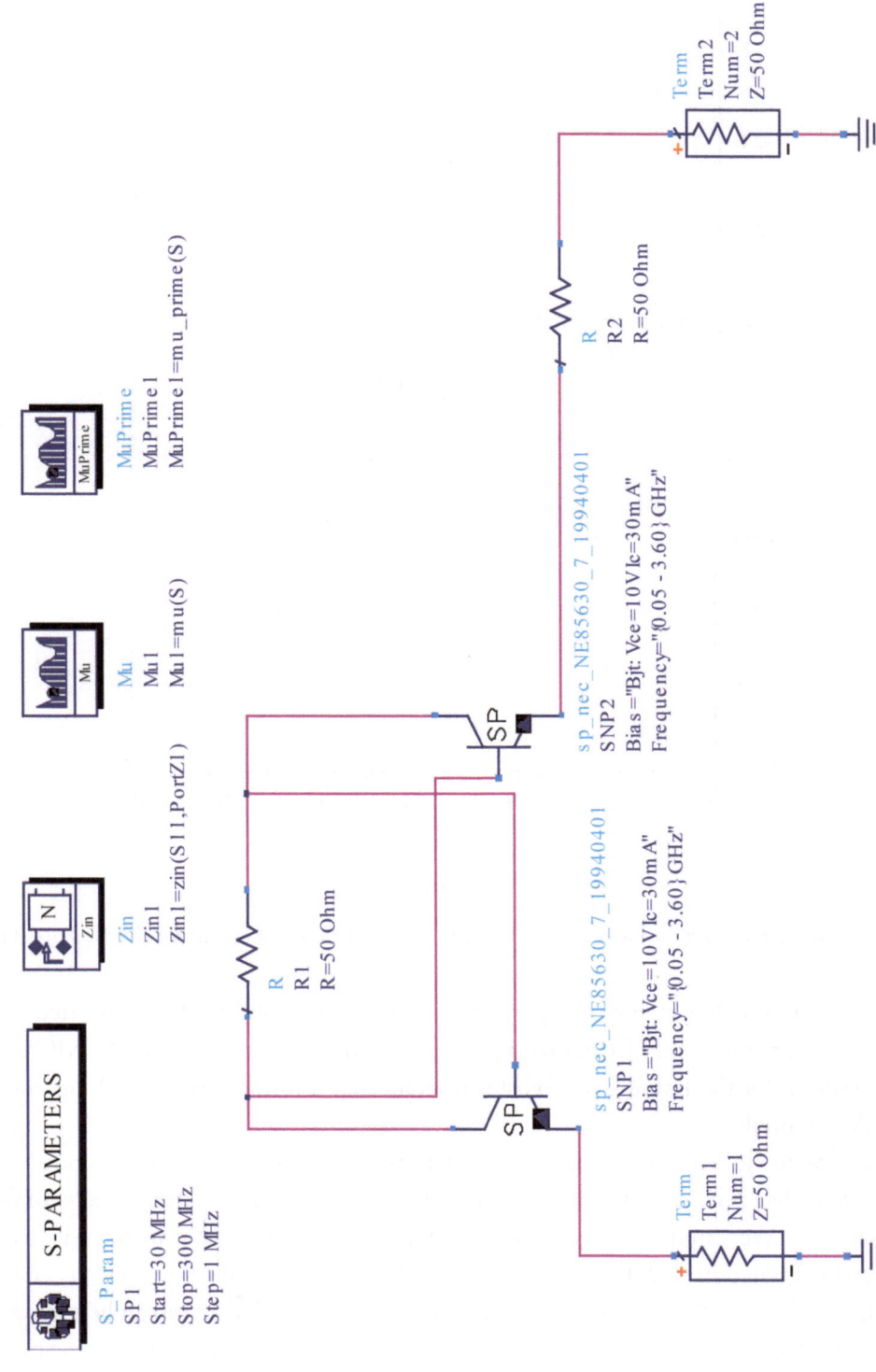

FIGURE 47: Schematic captured from Agilent ADS of the all-pass test circuit for the NE85630 FNR

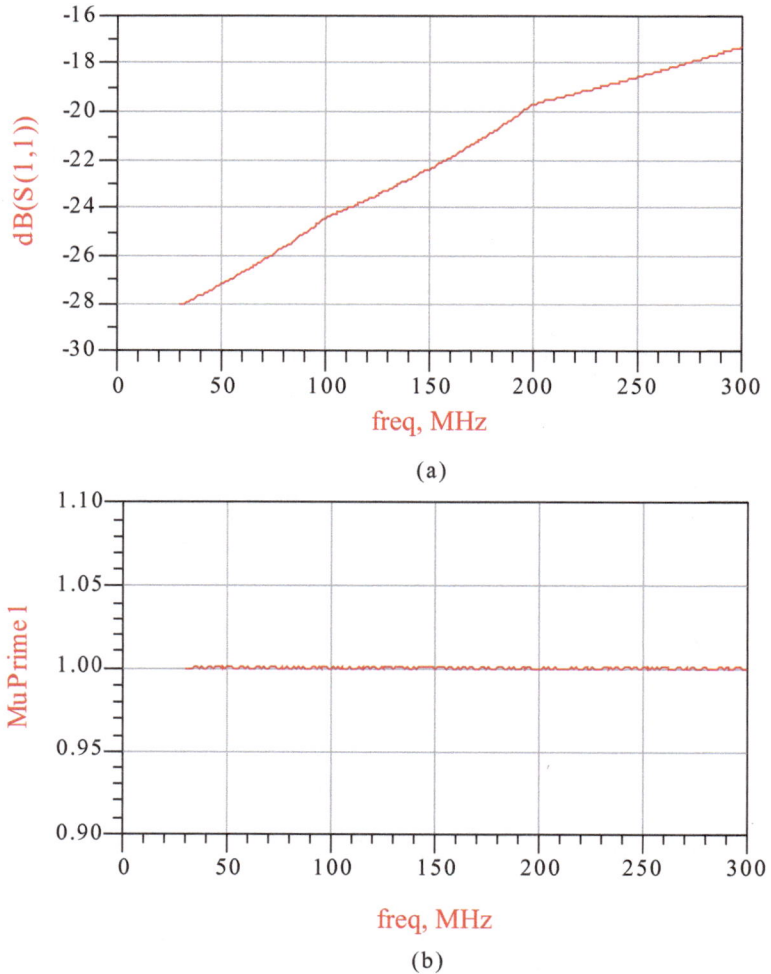

FIGURE 48: Simulated (a) return loss and (b) stability of the all-pass test circuit for the NE85630 FNR

neglected here, but do affect the circuit performance especially stability. The simulated results for the NE85630 are the best FNR results that we obtained. Thus, the NE85630 FNIC is used in the next section for the floating non-Foster reactance used in the active matching network for our ESA monopole.

In addition to the NIC circuits discussed in detail in this section, we also made considerable effort trying to realize NIC circuits that utilized CCII- blocks implemented as cascades of GaAs PHEMT devices. We simulated these circuits extensively and were able to obtain excellent performance in simulation with bandwidths greater than 1 GHz. Unfortunately, our attempts to physically implement these designs have all ended in failure. Other researchers have also reported a lack of success using this approach [16], and so we have abandoned it for the present.

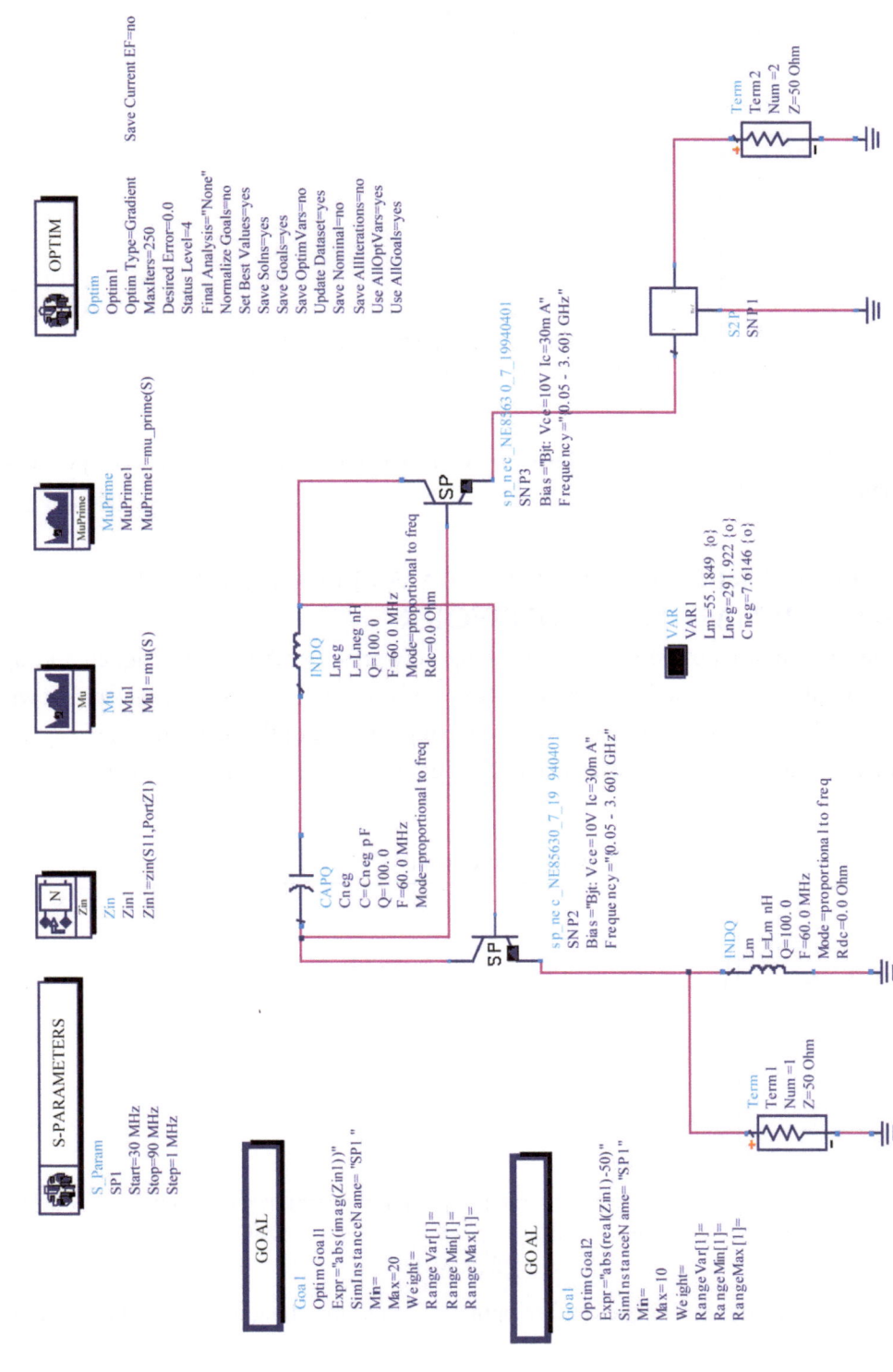

FIGURE 49: Schematic captured from Agilent ADS of VHF monopole with active matching network

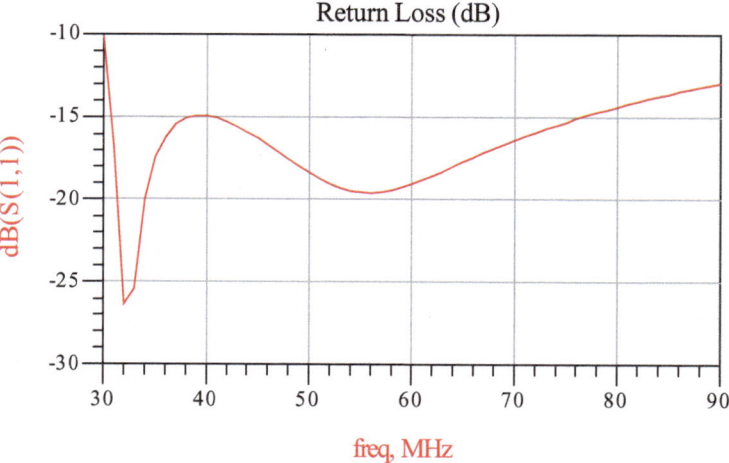

FIGURE 50: Return loss at input of optimized active matching network and antenna computed using Agilent ADS

SIMULATED PERFORMANCE OF ESA WITH A PRACTICAL NON-FOSTER MATCHING NETWORK

To illustrate the potential of non-Foster matching networks for ESAs, we designed and optimized in Agilent ADS a practical implementation of the active matching network shown in Fig. 12 for our ESA monopole antenna. We used a single FNIC of the form shown in Fig. 23 to implement the non-Foster series reactance consisting of $-(L_a + L_m)$ in series with $-C_a$. The

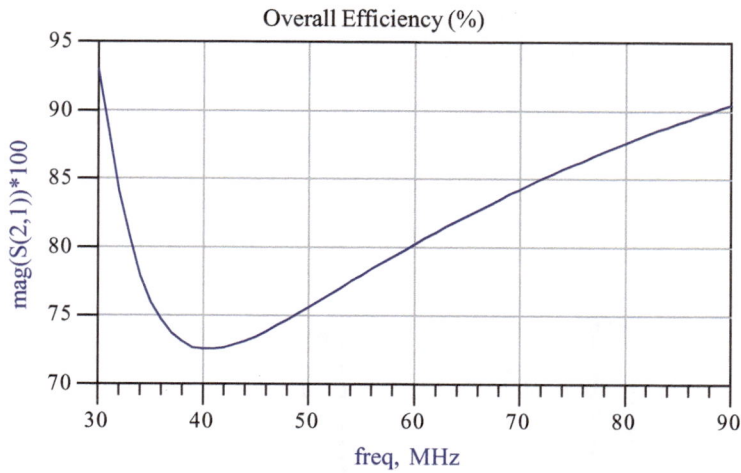

FIGURE 51: Overall efficiency (in percent) of optimized active matching network and antenna computed using Agilent ADS

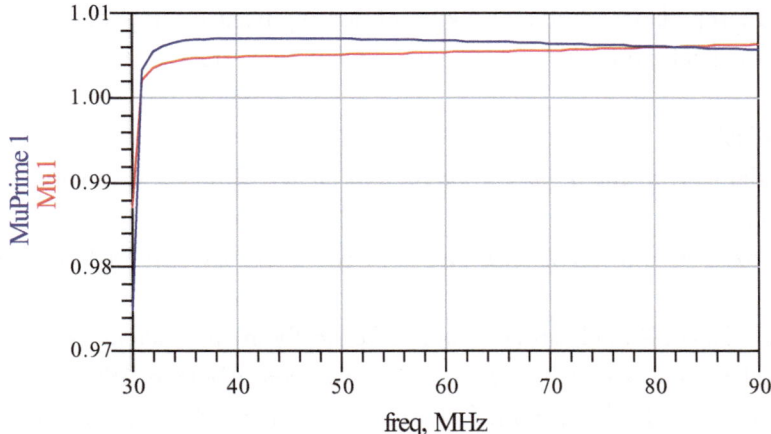

FIGURE 52: Small-signal geometrically derived stability factor for the optimized active matching network and antenna computed using Agilent ADS

active devices (NE85630 silicon bipolar NPN transistors) were modeled using the S-parameter library in Agilent ADS. Not surprisingly, we found that the simulated NIC performance was far from ideal. Nevertheless, using the gradient optimizer in Agilent ADS, we were able to adjust the values of the capacitor and inductors in the matching network to achieve remarkable broadband performance from the ESA monopole. The schematic of the two-port antenna model and active matching network captured from Agilent ADS is shown in Fig. 49. Note the presence of the measurement component for the small-signal geometrically derived stability factors μ and μ'. The computed return loss looking into the input of the matching network is shown in Fig. 50, and the total efficiency of the antenna together with the active matching network is shown in Fig. 51. Note that an extremely broadband and highly efficient match has been achieved. The geometrically derived stability factors as a function of frequency are shown in Fig. 52. These factors must be strictly greater than 1 for the circuit to be unconditionally stable. Note that below about 31 MHz, the overall circuit is not unconditionally stable. This situation should ultimately be remedied to avoid spurious radiation from the antenna.

CONCLUSIONS

In this lecture, we discussed an exciting new area of research in antenna technology, namely, the use of non-Foster circuit elements in the matching network of an electrically small antenna. The contributions of this lecture were to summarize the current state-of-the-art in this subject, and to introduce some new theoretical and practical tools for helping others to continue the advancement of this technology. The new contributions include a rigorous method for generating a two-port model for an antenna, an all-pass test circuit for evaluating the performance of floating negative impedances, and a new kind of floating negative impedance converter formed from two back-to-back grounded negative impedance converters.

REFERENCES

[1] C. A. Balanis, *Antenna Theory: Analysis and Design.* 3rd ed. New York: John Wiley and Sons, Inc., 2005.

[2] D. M. Pozar, *Microwave Engineering.* 3rd ed., New York: John Wiley and Sons, Inc., 2005.

[3] G. Skahill, R. M. Rudich, and J. Piero, "Electrically small, efficient, wide-band, low-noise antenna elements," *Antenna Applications Symposium*, Allerton, 1998.

[4] G. Skahill, R. M. Rudich, and J. A. Piero, "Apparatus and method for broadband matching of electrically small antennas," U.S. Patent Number 6,121,940, Sept. 19, 2000.

[5] J. L. Merill, "Theory of the negative impedance converter," *Bell Syst. Tech. J.*, Vol. 30, pp. 88–109, Jan. 1951.

[6] Yamaha, "Advanced YST," *Technology-Advanced YST* [Online]. Available: http://www.yamaha.com/yec/customer/technology/YST.htm [Accessed: Jan. 30, 2003].

[7] S. Dardillac, "Highly selective planar filter using negative resistances for loss compensation," *European Microwave Conference*, 2003, pp. 821–824.

[8] A. Antoniou, "Floating negative-impedance converters," *IEEE Trans. Circuit Theory (Corres.)*, Vol. CT-19, No. 2, pp. 209–212, Mar. 1972.

[9] S. E. Sussman-Fort, "Gyrator-based biquad filters and negative impedance converters for microwaves," *Int. J. RF Microwave CAE* Vol. 8, pp. 86–101, 1998.

[10] A. Sedra, G. Roberts, and F. Gohh, "The current conveyor: history, progress, and new results," *IEEE Proc. G*, Vol. 137, No. 2, pp. 78–87, Apr. 1990.

[11] Texas Instruments, *OPA690 Wideband Voltage-Feedback Operational Amplifier with Disable*, 2005.

[12] A. Sedra and K. C. Smith, *Microelectronic Circuits*, 4th ed., New York: Oxford University Press, 1998.

[13] Maxim, *MAX435/MAX436 Wideband Transconductance Amplifiers*, 1993.

[14] Texas Instruments, *THS3202 Low Distortion, 2 GHz, Current Feedback Amplifier*, 2004.

[15] S. E. Sussman-Fort, "Matching network design using non-Foster impedances," *IEEE Long Island Section, Circuits and Systems Society* [Online]. Available: http://www.ieee.li/cas/index.htm [Accessed: Dec. 6, 2005].

[16] S. E. Sussman-Fort and R. M. Rudish, "Progress in use of non-Foster impedances to match electrically-small antennas and arrays," *Antenna Applications Symposium*, Allerton, 2005.